Smart Electric and Hybrid Vehicles

In this book, recent developments, the future outlook, and advanced and analytical modeling techniques of smart electric and hybrid vehicles are explained with examples backed by experimental and numerical data. It also discusses the integration of newer developments like digital twin, artificial intelligence, nature-inspired algorithms, Internet of Things, and the role of Industry 4.0 in advancements in vehicle engineering. It compiles overall aspects of advancements in smart electric and hybrid vehicles by bringing the latest research and development by comprehensive range of mathematical, numerical, and simulation modeling, and management techniques to strengthen the engineering science and technological developments for the future.

Features:

- This book focuses on contemporary aspects of smart electric and hybrid vehicles techniques for new means and models for green environment.
- Discusses the role of artificial intelligence, machine learning, and machine vision tools in smart electric and hybrid vehicles.
- Presents design and analysis of charging stations and their sustainability roadmap for smart electric vehicles.
- Highlights the cyber and functional security of intelligent and hybrid vehicles.
- Explains diagnostics, prognostics, reliability, and durability issues in smart electric and hybrid vehicles.
- Covers the Internet of Things-based battery and charging management approach and effect of voltage drop in charging capacity of smart electric vehicles.

It is primarily written for senior undergraduates, graduate students, and academic researchers in the fields of electrical engineering, electronics and communication engineering, computer engineering, and automotive engineering.

Industrial and Manufacturing Systems and Technologies: Sustainable and Intelligent Perspectives

Series Editors: Ajay Kumar and Parveen Kumar

This book series provide a platform of single and multidisciplinary advance research on industrial and manufacturing engineering to introduce a requirement for a synergy from management and engineering. It provides the readers paradigms and systems of intelligence and sustainability in industrial and manufacturing engineering. It will contribute towards dissemination of 5.0 industrial revolution techniques towards academia, industry, and various practical communities involved in mechanical, industrial, computer science, automobile, electrical, electronics, aerospace engineering, supply chain, logistics, banking, finance management, circular economy, medical, pharmaceuticals, and healthcare industries. It will focus on important topics such as additive manufacturing, virtual reality in industrial engineering, industrial Internet of Things, intelligent engineering asset management system, and modeling of intelligent industrial engineering systems.

Smart Electric and Hybrid Vehicles
Fundamentals, Strategies and Applications
Edited by Ajay Kumar, D. K. Rajak, Parveen Kumar and Sarita Rathee

Smart Electric and Hybrid Vehicles
Design, Modeling, and Assessment by Industry 4.0 Approaches
Edited by Ajay Kumar, D. K. Rajak, Parveen Kumar and Sarita Rathee

For more information about this series, please visit: www.routledge.com/Industrial-and-Manufacturing-Systems-and-Technologies-Sustainable-and/book-series/CRCIAMSAT

Smart Electric and Hybrid Vehicles

Design, Modeling, and Assessment by Industry 4.0 Approaches

Edited by Ajay Kumar, D. K. Rajak,
Parveen Kumar, and Sarita Rathee

CRC Press
Taylor & Francis Group
Boca Raton London New York

CRC Press is an imprint of the
Taylor & Francis Group, an **informa** business

Designed cover image: shutterstock

First edition published 2024
by CRC Press
2385 NW Executive Center Drive, Suite 320, Boca Raton FL 33431

and by CRC Press
4 Park Square, Milton Park, Abingdon, Oxon, OX14 4RN

CRC Press is an imprint of Taylor & Francis Group, LLC

ISBN: 9781032801285 (hbk)
ISBN: 9781032820040 (pbk)
ISBN: 9781003502470 (ebk)

DOI: 10.1201/9781003502470

Typeset in Sabon
by Newgen Publishing UK

Contents

Preface

The electric vehicle market has been gradually gaining prominence in the world due to the rise in pollution levels caused by traditional IC engine-based vehicles. The advantages of electric vehicles are multi-pronged in terms of cost, energy efficiency, and environmental impact. The running and maintenance cost are considerably less than traditional models. The harmful exhaust emissions are reduced, besides the greenhouse gas emissions, when the electric vehicle is supplied from a renewable energy source. However, apart from some Western nations, many developing and underdeveloped countries have yet to take up this initiative. This lack of enthusiasm has been primarily attributed to the capital investment required for charging infrastructure and the slow transition of energy generation from the fossil fuel to the renewable energy format. Currently, there are very few charging stations, and the construction of the same needs to be ramped up to supplement the growth of electric vehicles. Grid integration issues also crop up when the electric vehicle is used to either do supply addition to or draw power from the grid. These problems need to be fixed by designing, modeling, optimizing, and by assessing the technology of smart electric and hybrid vehicles at all the levels to enhance the future of energy-efficient transportation. This book discusses various design, modeling, and optimization strategies for smart vehicles like Internet of Things-based battery and charging management approach, comparative analysis and optimization of barriers and enables smart and hybrid vehicles by fuzzy, interpretive structural modeling strategies, Sustainable experimental, numerical, simulation, and computational approaches, the role of artificial intelligence, machine learning and machine vision tools, optimization of charging criteria, power management, and battery degradation; and assessment of smart electric and hybrid vehicles such as government regulations, policies, tradeoffs, market prospects, competition, compliance, commercialization, standardization, and demonstration projects of smart electric and hybrid vehicles, Situational awareness, sensing control, road trajectory prediction, and navigation systems design, analysis of cost-effectiveness, on road performance and life cycle assessment cyber and functional security, efficiency, performance and parameters mapping, conductive electrical interface and protection against electric shock, effect of voltage drop in charging capacity, diagnostics, prognostics, and reliability and durability issues, psychology, challenges, recommendations for smart and electric vehicles.

This book is of interest for individuals in the automotive sector and allied industries, policymakers, practitioners, engineers, technicians, researchers, academicians, and students looking for updated information on the technology, economics, policy, and environmental aspects of electric vehicles.

The book consists of 11 chapters that describe perspectives of smart electric and hybrid vehicles. Chapter 1 'Smart electric and hybrid vehicles architecture' focuses on the architecture of smart and hybrid electric vehicle and their operation. This chapter also covers types of different types of batteries and configuration of hybrid electric vehicle using conventional and non-conventional energy sources. Chapter 2 'Proof of concept: an on-road recharging e-vehicle' proposed a system that equipped a transmitter coil that obtains a power source at the bottom of the road and the receiver coil will be placed at the bottom of the EV prototype. Chapter 3 'Cylindrical lithium-ion battery: fundamentals, design, and manufacturing processes' seeks to provide researchers with clarity on the fundamentals, design, and manufacturing processes of LiBs. Chapter 4 'Revolutionizing vehicular braking: the uni-booster integrated disc brake system' emphasizes the significance of designing and developing a uni-booster integrated disc brake system for the manufacturing of LCVs. Chapter 5 'Analyzing the impact of bumper height on pedestrian injuries using explicit dynamics' demonstrated that pedestrian injury types and severity are significantly influenced by bumper height. Chapter 6 'Improvise in sigma level in automobile industries' deals with process improvement methods to enhance productivity, reduce defects, ease the methodology, and save money. Chapter 7 'Navigating the roadmap of intelligent transport in Indian smart cities: sectorial analysis, challenges, and key elements of data science enabled solutions' explain real-time datasets in ITS-related applications, such as forecasting passenger flow, estimating travel and stay time, developing optimal routes, planning efficient networks, and providing reliable services at low operational costs. Chapter 8 'Enhancing aircraft inspection through computer vision: mitigating human error and improving safety' presents computer vision as an effective tool that not only enhances the reliability of aircraft inspections but also contributes to a safer aviation industry. Chapter 9 'Unleashing the power of YOLOv5: revolutionizing person detection and counting in restricted zones' examines the application of YOLOv5 for person identification in crowds. Chapter 10 'Revolutionizing cloud computing security with the power of artificial intelligence: advancements, challenges, and opportunities' discusses the challenges and limitations of AI-based cloud security solutions and highlight some of the key research directions that could help overcome these challenges. Chapter 11 'HDVF: hazy object detection network based on variational features' proposed a hazy object detection network based on variational features, which is capable of dealing with haze in the images for feature extraction.

This book is intended for both academia and industry. The postgraduate students, Ph.D. students and researchers in universities and institutions, who are involved in the areas of smart electric and hybrid vehicles, will find this compilation useful.

The editors acknowledge the professional support received from CRC Press and express their gratitude for this opportunity.

Reader's observations, suggestions, and queries are welcome.

Ajay Kumar
D.K. Rajak
Parveen Kumar
Sarita Rathee

Editors

Ajay Kumar is currently a Professor in the Mechanical Engineering Department, School of Engineering and Technology, JECRC University, Jaipur, Rajasthan, India. He earned his Ph.D. in advanced manufacturing from Guru Jambheshwar University of Science & Technology, Hisar, India after B.Tech. (Hons.) and M.Tech. (Distinction) from Maharshi Dayanand University, Rohtak, India. His areas of research include artificial intelligence, intelligent manufacturing, incremental sheet forming, additive manufacturing, intelligent vehicles, advanced manufacturing, Industry 4.0, waste management, and optimization techniques. He has over 60 publications in international journals of repute including SCOPUS, Web of Science and SCI indexed database and refereed international conferences. He has co-authored/co-edited several books including the following:

- *Incremental Sheet Forming Technologies: Principles, merits, limitations, and applications*, CRC Press, Taylor and Francis, ISBN 9780367276744, https://doi.org/10.1201/9780429298905
- Managing Guest Editor of Journal Proceedings, Ajay Kumar, Journal of Physics: Conference Series, Volume 1950, International Conference on Mechatronics and Artificial Intelligence (ICMAI) 2021 27 February 2021, Gurgaon, India, https://iopscience.iop.org/issue/1742-6596/1950/1
- *Advancements in Additive Manufacturing: Artificial Intelligence, Nature Inspired and Bio-manufacturing*, ISBN: 9780323918343, ELSEVIER, https://doi.org/10.1016/C2020-0-03877-6
- *Handbook of Sustainable Materials: Modelling, Characterization, and Optimization*, CRC Press, Taylor and Francis, ISBN: 978-1-032-28632-7, https://doi.org/10.1201/9781003297772
- *Waste Recovery and Management: An Approach Towards Sustainable Development Goals*, CRC Press, Taylor and Francis, ISBN: 9781032281933, https://doi.org/10.1201/9781003359784

- *Smart Manufacturing: Forecasting the future of Industry 4.0*, CRC Press, Taylor and Francis, ISBN: 9781032363431, https://doi.org/10.1201/9781003333760
- *Handbook of Flexible and Smart Sheet Forming Techniques: Industry 4.0 Perspectives*, ISBN: 9781119986409, WILEY. https://onlinelibrary.wiley.com/doi/book/10.1002/9781119986454
- *Modeling, Characterization, and Processing of Smart Materials*, IGI Global, and ISBN: 9781668492246. DOI: 10.4018/978-1-6684-9224-6

He has organized various national and international events including an International Conference on Mechatronics and Artificial Intelligence (ICMAI-2021) as conference chair. He has more than 15 national and international patents in his credit. He has supervised more than eight M.Tech, Ph.D scholars and numerous undergraduate projects/theses. He has a total of 13 years of experience in teaching and research. He is a guest editor and review editor of reputed journals including *Frontiers in Sustainability.*

D.K. Rajak is a scientist at CSIR-Advanced Materials and Processes Research Institute, Bhopal, Madhya Pradesh, India. He completed his Ph.D. from the Indian Institute of Technology (ISM), Dhanbad, Jharkhand. He received many prestigious awards and was recognized in the World Top 2% Scientists by Sanford University in 2022 & 2023. He has published over 100 peer-reviewed journal papers and has three books, 16 book chapters, 18 patents, two designs, and one copyright. His recent work involves developing lightweight alloy composite foams and composites for applications requiring low density and high specific energy absorption in automobile, aerospace, marine, noise, and thermal management sectors.

Parveen Kumar is currently an Assistant Professor and Head in the Department of Mechanical Engineering at Rawal Institute of Engineering and Technology, Faridabad, Haryana, India. Currently, he is pursuing a Ph.D. from the National Institute of Technology, Kurukshetra, Haryana, India. He completed his B.Tech. (Hons.) from Kurukshetra University, Kurukshetra, India and M.Tech. (Distinction) in Manufacturing and Automation from Maharshi Dayanand University, Rohtak, India. His areas of research include intelligent manufacturing systems, materials; die less forming, design of automotive systems, additive manufacturing, CAD/CAM and artificial Intelligence, machine learning and Internet of Things in manufacturing, multi-objective optimization techniques. He has over 20 publications in international journals of repute including SCOPUS, Web of Science and SCI indexed database and refereed international conferences. He has eight national and international patents in his credit. He has supervised more than ten M.Tech. scholars and numerous undergraduate projects/

theses. He has a total of 12 years of experience in teaching and research. He has co-authored/co-edited several books including the following:

- *Waste Recovery and Management: An Approach Towards Sustainable Development Goals*, CRC Press, Taylor and Francis, ISBN: 9781032281933, https://doi.org/10.1201/9781003359784
- *Handbook of Sustainable Materials: Introduction, Modelling, Characterization and Optimization*, CRC Press, Taylor and Francis, ISBN: 9781032295874, https://doi.org/10.1201/9781003297772
- *Handbook of Smart Manufacturing: Forecasting the future of Industry 4.0*, CRC Press, Taylor and Francis, ISBN: 9781032363431, https://doi.org/10.1201/9781003333760
- *Handbook of Flexible and Smart Sheet Forming Techniques: Industry 4.0 Perspectives*, ISBN: 9781119986409, WILEY. https://onlinelibrary.wiley.com/doi/book/10.1002/9781119986454
- *Modeling, Characterization, and Processing of Smart Materials*, IGI Global, and ISBN: 9781668492246. DOI: 10.4018/978-1-6684-9224-6

He has organized a series of Faculty Development Programmes, workshops, and seminars for researchers and UG level students.

Sarita Rathee is an Assistant Professor in the Department of Electrical and Electronics Engineering, School of Engineering and Technology, JECRC University, Jaipur, Rajasthan, India. She earned her M.Tech (Distinction) in electrical and electronics engineering from Maharshi Dayanand University, Rohtak, India after B.Tech. (Hons.) in electrical engineering. Her areas of research include artificial intelligence, IOT, cloud computing, Industry 4.0, waste management, and optimization techniques. She has over 10 publications in international journals of repute including SCOPUS, Web of Science and SCI indexed database and refereed international conferences. She has more than five national and international patents to her credit. She has supervised more than four M.Tech scholars and numerous undergraduate projects/theses. She has a total of five years of experience in teaching and research. She has won several proficiency awards during her career, including merit awards, best teacher awards, and so on. She is co-adviser of QCFI, Delhi Chapter student cell at JECRC University and has also authored many in-house course notes, lab manuals, monographs, and invited chapters in books. She has attended a series of Faculty Development Programs, International Conferences, workshops, and seminars for researchers, Ph.D, UG, and PG level students. She teaches the following courses at the graduate and postgraduate level: instrumentation engineering, IOT, FACTS devices, power electronics, electric vehicles and so on. She is associated with many research, academic, and professional societies in various capacities.

Contributors

Abhishek Agarwal
Department of Mechanical Engineering, College of Science & Technology, Royal University of Bhutan, Chukka, Bhutan

Samir B. Patel
Department of Computer Engineering, Pandit Deendayal Energy University, Gandhingar, India

Vineet Bhatia
Defence Institute of Advanced Technology, Pune, India

Rafael Cavicchioli Batista
Department of Mechanical & Manufacturing Engineering Technology, Vermont State University, Vermont, USA

Sachin Chaudhary
School of Computer Science and Applications, IIMT University, Meerut, India

Vishal Dabra
Mechanical Engineering Department, P.I.E.T, Samalkha, India

Gurumukh Das
Department of Mechanical Engineering, Faculty of Engineering, Dayalbagh Educational Institute (Deemed University), Agra, India

Rajeev Gupta
Department of Computer Engineering, Pandit Deendayal Energy University, Gandhingar, India

Adash Gurung
Department of Mechanical Engineering, College of Science & Technology, Royal University of Bhutan, Chukka, Bhutan

Suriya Prakash J.
Department of Computer Science and Engineering, JAIN (Deemed-to-be-University), Bangaluru, India

Kewal Krishan Sharma
School of Computer Science and Applications, IIMT University, Meerut, India

Amit Kumar Dubey
Mechanical Engineering Department, P.I.E.T, Samalkha, India

Tarun Kumar Vashishth
School of Computer Science and Applications, IIMT University, Meerut, India

Achint Kumar
Department of Mechanical Engineering, Faculty of Engineering, Dayalbagh Educational Institute (Deemed University), Agra, India

Parveen Kumar
Department of Mechanical Engineering, Rawal Institute of Engineering and Technology, Faridabad, India

Ajay Kumar
Department of Mechanical Engineering, School of Engineering and Technology, JECRC University, Jaipur, India

Aman Kumar
Department of Mechanical Engineering, UIET, MDU, Rohtak, India

Sachin Kumar
Department of Mechanical Engineering, UIET, MDU, Rohtak, India

Bhupendra Kumar
School of Computer Science and Applications, IIMT University, Meerut, India

Akhil Nigam
Electrical Engineering Department, Chandigarh University, Mohali, India

Neethu P.S.
Department of Electronics and Communication Engineering, School of Engineering and Technology, Christ (Deemed-to-be-University), Bangaluru, India

Rajneesh Panwar
School of Computer Science and Applications, IIMT University, Meerut, India

Madhuri Patel
Department of Computer Engineering, Pandit Deendayal Energy University, Gandhingar, India

Suguna R.
Department of Computer Science and Engineering, Vel Tech Rangarajan Dr. Sagunthala R&D Institute of Science and Technology, Avadi, Chennai, India

Vikas Sharma
School of Computer Science and Applications, IIMT University, Meerut, India

Sumati Sidharth
Defence Institute of Advanced Technology, Pune, India

Anil Singh Parihar
Department of Computer Science & Engineering, Delhi Technological University, New Delhi, India

Kavinder Singh
Department of Computer Science & Engineering, Delhi Technological University, New Delhi, India

Debabrata Swain
Department of Computer Engineering, Pandit Deendayal Energy University, Gandhingar, India

Ching Yee Yong
School of Engineering and Technology, University of Technology Sarawak, No. 1, Jalan Universiti, Sibu, Sarawak, Malaysia

Chuan Zheng Ling
School of Engineering and Technology, University of Technology Sarawak, No. 1, Jalan Universiti, Sibu, Sarawak, Malaysia

Acknowledgments

The editors are grateful to CRC Press for publishing this book in the buzz area of *Smart Electric and Hybrid Vehicles: Design, Modeling, and Assessment by Industry 4.0 Approaches*. The editors express gratitude to Mr. Gauravjeet Singh Reen, Executive Editor, CRC Press, for giving consent to publish this work. The editors also extend their gratitude to Ms. Mehnaz Hussain, CRC Press, for her support.

The editors, wish to thank all the chapter authors for contributing their valuable research and experience to compile this volume. The chapter authors, corresponding authors in particular, deserve special acknowledgements for bearing with the editors, who persistently kept bothering them for deadlines, and with their remarks.

Dr. Ajay also wishes to express his gratitude to his parents, Sh. Jagdish and Smt. Kamla, and his loving brother Sh. Parveen for their true and endless support. They have made him able to walk tall before the world regardless of sacrificing their happiness and living in a small village. He cannot close these prefatory remarks without expressing his deep sense of gratitude and reverence to his life partner Mrs. Sarita Rathee for her understanding, care, support, and encouragement to keep his moral high all the time. No magnitude of words can ever quantify the love and gratitude he feels in thanking his daughters, Sejal Rathee and Mahi Rathee, and son Kushal Rathee who are the world's best children.

Finally, the editors dedicate this work to the divine creator and express their indebtedness to the 'ALMIGHTY' for gifting them power to yield their ideas and concepts into substantial manifestation. The editors believe that this book will enlighten readers about each feature and characteristic of smart electric and hybrid vehicles.

Ajay Kumar
D.K. Rajak
Parveen Kumar
Sarita Rathee

Chapter 1

Smart electric and hybrid vehicle architecture

Akhil Nigam,[1] Parveen Kumar,[2] and Ajay Kumar[3]

[1]Electrical Engineering Department, Chandigarh University, Mohali, India
[2]Department of Mechanical Engineering, Rawal Institute of Engineering and Technology, Faridabad, India
[3]Department of Mechanical Engineering, School of Engineering and Technology, JECRC University, Jaipur, India

1.1 INTRODUCTION

Increasing greenhouse gas emissions has increased the temperature on Earth over the last few years. Various sectors are contributing to this, such as the transportation sector, which produces around 25% of the total greenhouse gas emission. Road vehicles are responsible for the production of around 75% carbon dioxide gas emission. Hence a step has been taken in the transportation field to introduce electric-based vehicles in the transportation sector [1]. In the United States of America, production of CO_2 emission was around 4.7 billion tons, which is low compared to previous years and is due to the COVID-19 pandemic. Along with this, light- and medium-sized vehicles contribute around 80% gas emission as per the report by the United States Environmental Protection Agency. Many government agencies have started work and have transition goals to overcome the problem of greenhouse gas emission. But there are also various challenges in this scenario in addition to the electrification of commercial vehicles and providing awareness among people. Initially, design preparation and configuration are the most important parts and then battery installation is needed for charging purposes. For this, many studies have been performeed by authors and researchers and many works towards this goal have been started. Along with the electrification of commercial vehicles, researchers have also focused on their hybridization for making with advanced emerging technologies [2]. Hybridization is required due to lack of sufficient infrastructure, small size of batteries, and discontinuity of the availability of fuels. Having only a single source may present a problem for driving electric vehicles in hybrid mode.

These electric vehicles can protect environment efficiently by reduction of gas emission. These days various forms of electric vehicles are available due to the presence of different types of batteries and charging facilities. Electric vehicles or other forms of EVs are considered for a next-generation transportation system. People are helping to develop this facility butthe cost

DOI: 10.1201/9781003502470-1

of EVs is high due to the non-availability of charging stations and maintenance facilities. By 2030 electric vehicles will be considered only in place of conventional fuel-based vehicles. For better transportation, conventional vehicles have to be replaced to achieve a greater gain and by the presence of an energy storage system it can maintain high-power capabilities with safety measures in the next few years [3]. Various batteries are available, such as nickel metal and lithium ion, as they have higher energy density as compared to lead acid batteries. Most of the types of lithium-ion battery are preferred as they are more affordable and quickly reachable. These are also available at low cost and provide a more efficient response. For energy provision, sometimes fuel cells can also be used but their manufacturing cost is high. Hence flywheel-type storage can also be used as it has large capacity rather than a capacitor. On the other hand, a capacitor can deliver fast energy and it has higher potential energy than a flywheel [4]. Most of the times, a lithium-ion battery is used, which was discovered in the early 1990s. In this configuration, lithium as the cathode and carbon as the anode can be influenced for the generation of electricity. These are the least in cost and size. Such types of batteries can be recycled with a minimum amount of wastage. Many authors have proposed configuration of electric vehicles by the combination of an ICE engine with one or more motor, which are used for storage in batteries. Further, electric vehicles have been configured with a novel approach by many researchers to identify hybrid topology.

1.1.1 Description of components

There are various components present in electric vehicles, which take part in the conduction of electric vehicles. In addition to the conduction of electricity, it can also be utilized for energy storage. Below, we will discuss the components of electric vehicles:

(A) **Battery pack:** it is a device that is used to retain charge in itself. This is used for traction of electric vehicles and it can be used to store and deliver DC supply to the inverter. Hence electricity generated can be used to drive electric vehicles.
(B) **Electric motor:** it is considered as a secondary energy source for assisting the engine and it provides real power. It is used to drive wheels and nowadays a brushless-type motor is preferred.
(C) **Power inverter:** it is used to convert DC power into AC power. It also uses a regenerative braking system for the generation of AC. Further it can be used to recharge batteries.
(D) **DC-DC converter:** it is used as a complementary part of electric vehicles, which is used to convert high voltage to low voltage. Sometimes low power can be required by the components. It can be

used as a medium for battery charging, which maintains voltage and current.

(E) **Charging port:** it is another important component for charging batteries through an external source of supply. The location of the charging port can be in front or the back side of the vehicle.

(F) **Controller:** it is used as a power controller unit in electric vehicles for monitoring power delivery to the motor using a battery and inverter. It regulates electric energy from the battery to motor.

(G) **Cooling system:** cooling is provided by a proper thermal management system as heat generated by the electric vehicle needs to be utilized. Some methods are available such as liquid cooling, forced air cooling, and thermoelectric cooling.

1.2 TYPES OF ELECTRIC VEHICLES

Based on the configuration of electric vehicles it has been classified into the following types as given below.

1.2.1 Battery electric vehicle

This type of vehicle does not require an IC engine it is driven by a battery and electric drive train. Such a type of EV contains drive train, motor, inverter, battery, and control module. DC power is altered into AC power. A controller receives a signal from the pedal accelerator it modifies the frequency of AC produced from the inverter to motor for regulating the speed of motor. The motor is then connected to the wheel, which rotates it by the help of a gear. Whenever brakes are applied or the speed of the vehicle reduces, the motor converts into an alternator and produces electricity, which then returns to the battery for storage purpose. The block diagram of a battery electric vehicle is shown in Figure 1.1.

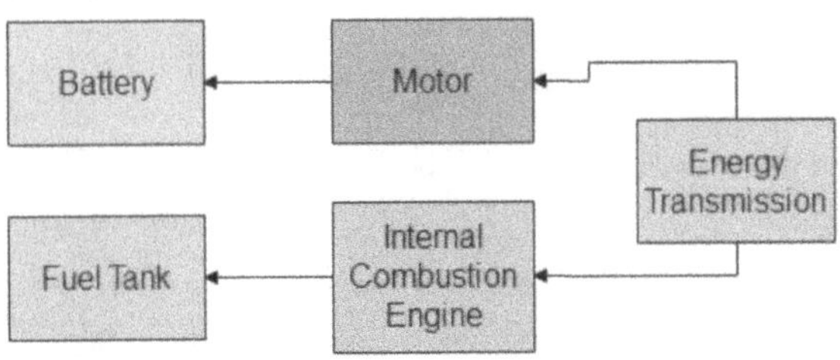

Figure 1.1 Battery electric vehicle.

1.2.2 Hybrid electric vehicle

This type of vehicle contains configuration in series, parallel, or both in series-parallel mode. This EV is operated by combination of both ICE and motor. The battery is used to give power to the motor and the ICE is driven by gasoline fuel or another type of fuel. The combination of the ICE and motor rotates the gear box, which runs the wheel. The main components are the ICE, motor, battery pack, fuel tank, and control module. The block diagram of a hybrid electric vehicle is shown in Figure 1.2.

1.2.3 Plug-in hybrid electric vehicle

This type of motor is preferred as a series motor, which combines both electric motorand ICE. This vehicle can be operated on conventional fuel or alternative fuel and by a rechargeable battery system. The battery can be charged through an electric outlet or EV charging station. Such a vehicle contains an ICE, motor, inverter, battery, battery charger, and control module. The block diagram of a plug-in hybrid electric vehicle is shown in Figure 1.3.

1.2.4 Fuel cell electric vehicle

These types of electric vehicles use fuel cell technology and sometimes it is also called a fuel cell vehicle. It produces the electricity required to drive the vehicle by itself. The working of this type of electric vehicle is different from a plug-in electric vehicle. The main components are electric motor, fuel cell stack, battery, converter, controller, and hydrogen storage tank. The block diagram of a fuel cell electric vehicle is shown in Figure 1.4.

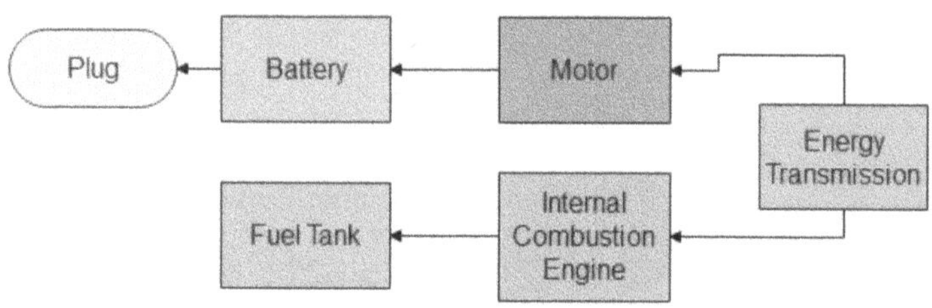

Figure 1.2 Hybrid electric vehicle.

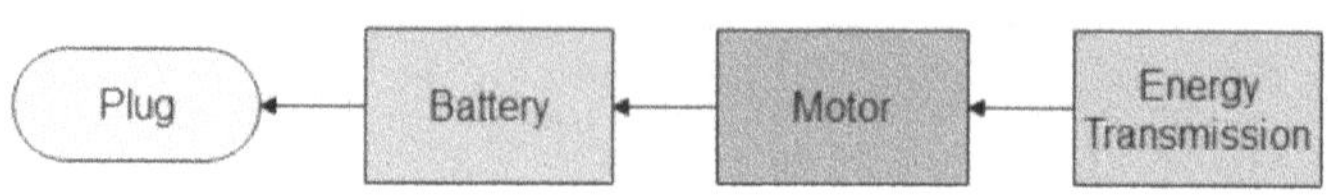

Figure 1.3 Plug-in hybrid electric vehicle.

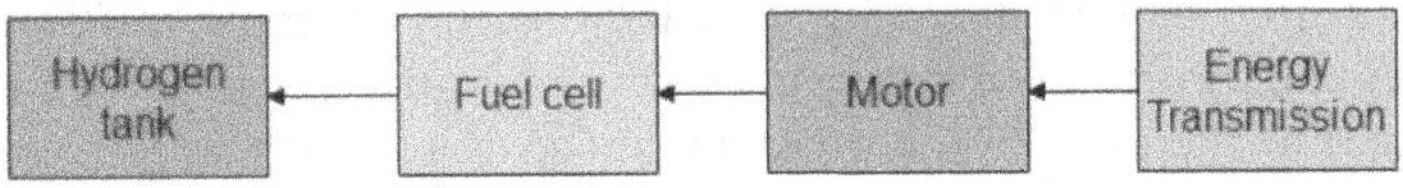

Figure 1.4 Fuel cell electric vehicle.

Table 1.1 Difference between electric vehicle and hybrid vehicle

Aspect	*Electric vehicle*	*Hybrid vehicle*
Primary energy source	Electricity	Diesel fuel or gasoline fuel
Type of engine	Electric motor	Combination of ICE and electric motor
Cost	Low cost	High cost
Level of emission	Zero emission level	Low emission level
Need of charging	It requires	It does not require
Driving range	It has low driving range	It has high driving range

1.3 MERITS AND DEMERITS OF A HYBRID VEHICLE

1.3.1 Merits

1. It is more efficient to use.
2. It is environmentally friendly.
3. Easy implementation with other energy source.
4. Require less maintenance.

1.3.2 Demerits

1. It may be expensive.
2. Requires skilled people.
3. Probability of the production of less power.

1.4 DIFFERENCE BETWEEN ELECTRIC VEHICLE & HYBRID VEHICLE

Table 1.1 shows the difference between electric and hybrid vehicles in different aspects for selecting at appropriate conditions.

1.5 BATTERY TECHNOLOGY AND ENERGY STORAGE SYSTEM

These are preferred as a rechargeable energy storage system, which stores energy from solar PV cells or the grid then delivers it for household applications. With sunlight, the battery energy storage system is charged by alternate electricity produced by solar light. Using emerging software technologies and algorithms it is easy to optimize for attaining improved energy. After that at the time of energy requirement, the battery storage provides energy.

These have wide commercial applications, such as peak shaving, emergency grid, load shifting, grid service, etc. Energy storage is another important aspect for enhancing power quality, matching energy demand etc. These batteries are one of the cheapest usage applications for numerous applications. Instead of using a super-capacitor, the hybrid storage system is also preferred, which leads to reduction in cost, weight, and volume of the storage system [5]. Batteries are incorporated into numerous groups, such as primary cell, secondary cell, fuel cell, and reserve cell. So battery storage technology is more suitable for renewable energy sources in an efficient way as environmental effects can be overcome by using these technologies. Most types of batteries like lithium ion (Li ion) and lead acid batteries are preferred for energy repository. Lead acid is an easy type of battery to use due to its simple implementation and maintenance but its operation depends upon temperature. The cost is very low and it can easily handle recycling. The same as lithium-ion, it is one of the widely used types of battery. Today's applications, such as mobiles, laptops, and tablets, are powered by Li-ion batteries. These types of batteries contain advantages of long life and low weight.

1.6 INFRASTRUCTURE OF CHARGING STATIONS (TYPES OF CHARGING)

For an electric vehicle, charging is another important factor because it depends upon the availability of charging stations located near consumers and their awareness to utilize them. Basically, dependency on the infrastructure of the EV systems' charging period also varies so accordingly it can be used. Based on power supply, there are two types of battery chargers, such as AC and DC chargers.

1.6.1 AC charger

These types of chargers perform slowly as they take at least 6–8 h for charging to full capacity.

1.6.2 DC charger

As a comparison to an AC charger they do fast charging. It has the ability to charge 85% typically in about 25–35 min.

EV systems can be charged through electric vehicle service interfacing at different charging speed; they are categorized as level-1, level-2 and DC fast charging. A level-1 type charger is used only for household applications. This charger requires much less maintenance and it can be charged overnight especially for plug-in electric vehicles. A level-2 charger requires higher rate equipments and these are preferred for public usage. For quick charging this type of charger can be used. For PHEVs it takes 1–2 h and for BEVs it takes 4–10 h. DC fast charging (DCFC) type chargers can charge accordingly 80% in 20–30 min. Longer distance vehicles use this type of charger. These are highly powered chargers and require the most equipped components. All electric vehicles cannot support this type of charger.

As there are various types of electric charging, such as public charging, residential charging, park charging, etc. Based upon requirement such as land, availability of batteries, awareness to people, low cost, maintenance cost, etc., must be considered. In residential charging stations, it can be operated by drawing less current from the grid to an extent by which demand can be overcome at the time of peak hours. In the case of parking charging stations, it can be functioned by proper installation of charging stations at shopping malls, restaurants, libraries, etc. There are some existing parking stations, which are now converted into smart charging stations with the access of the internet and WIFI-service. For public charging stations, it uses various topologies for fast charging services to the people through which it can boost the timing of batteries. Basically, a charging station carries a AC-DC converter at the front side and a DC-DC converter at the back side. Both types of converters are connected through a common DC link capacitor. These perform high switching operation thus high-frequency rectifiers are configured from insulated-gate bipolar transistor (IGBT) and metal oxide semiconductor field effect transistor (MOSFET). These types of transistors perform on the principle of pulse width modulation. Through these configurations it is easy to attain a more efficient and precise response [6].

1.7 VEHICLE TO GRID TECHNOLOGY AND SMART CHARGING STRATEGIES

Basically vehicle-to-grid technology or in short V2G technology is a technique in which energy is transferred back to the grid from the battery of an electric vehicle. If this EV battery can be discharged depends upon energy generation and energy consumption. This technology is becoming more influential due to the presence of an advanced level of innovation from the side of the source, switching-based power electronic devices, battery storage, and control strategies. By year 2030, the top EV hosting countries like the USA and China, peak demand will be around 650 GW. So according to the energy report, EVs may impact electricity regulation, control, and transportation and AI technology. The V2G technology has been implemented for improving the performance of power system with the help of recent advancement of the smart grid and microgrid. V2G technology utilizes EV as energy supplier and toggler in the energy grid means it can serve both as the source and load. Further it can be used as an alternate energy source by drawing unutilized energy from EV into the power grid, employing stability and reliability. In growing countries, EVs can be used as an alternate energy source where power is more desired at lower cost. If there is unavailability of solar energy, V2G technology helps to generate power plants, which provides reactive power compensation, load balancing, and active power monitoring, etc.

Using V2G technology provides a balancing of power flow between EV and the grid and is monitored to establish economic operation while reducing CO_2 gas emission and enhancing a better response. Further it can be divided into two types, such as unidirectional and bidirectional. In unidirectional type, energy transfers in a single direction from the grid to EV whereas bidirectional type, energy transfers in both directions hence this requires additional energy converters, which can perform in both charging and discharging mode [7].

1.8 ROLE OF EMERGING TECHNOLOGIES IN ELECTRIC VEHICLE

In the last few years, smart cities, smart grid, electric vehicles, and energy storage has become a vital part in the field of electricity and through the emerging technologies, production of electricity is ongoing towards reliability enhancement. So with the help of alternate or advanced technology people can attain sustainable energy from the power grid for optimal utilization of a smart grid and smart cities. In this current era, some emerging technologies have been developed, which can improve the performance of the EV system and can be accessible to each electrical sector.

1.8.1 Wireless charging technology

With the help of the principle of an induction charging system this system has been tuned with hardware devices and positioned at a precise angle, electric vehicles can be charged through the proper parking at efficient gas space. Through this, it can be operated in megahertz frequency bandwidth [8]. It provides specific convenience to owners of vehicles and provides reliable parking as compared to conventional types of parking. By employing this method, maintenance cost can be optimized. By proposed energy-driven technology demand and supply for electric vehicle can be increased. From the point of energy market, wireless charging rate has been increased with 49% percentage from year 2020–2025. Next WiTricity has been introduced for wireless charging purpose based on the principle of magnetic resonance. It can provide output power range from 4 kW to 11 kW and hence it can fulfill the requirement of charging using a PHEV system. Similarly Qualcomm Halo is another wireless charging system, which performs on high power application. It can cover up to 22 kW charging facility using a charging mat mounted on the floor. Next Zonecharge was introduced in 2016; it can be used for charging of electric vehicles up to a capacity of 6.6 kW and for buses it can be extended up to 30 kW. For achieving stable and reliable operation, an electric vehicle system may be configured in terms of load or energy source employed with a renewable or non-renewable energy system. This configuration helps to study electric vehicles for analysis of stability and feasibility with the existing system [9].

1.9 CONCLUSION

In this book chapter, the study of smart and hybrid electric vehicles has been done through the previous work. From the point of achieving efficient and reliable energy, it is necessary to use emerging techniques. By employing smart meters and smart sensors, electric vehicles can be operated in an autonomous mode. Providing availability of fast charging and batteries with energy storage these make EV system as enhanced emerged system. Through the combination of various components into the EV system it is easy to design and implement for producing electricity. In this chapter, types of different charging and their strategies have been employed for the generation of electricity.

1.10 FUTURE SCOPE OF ELECTRIC VEHICLE

So demand of the EV system is optimizing day by day through using emerging techniques and reduction of CO_2 gas emission. This is useful for the power supply system, through this energy can be supplied and produced as per required by users. Up to 2030, slow chargers will be utilized for charging of the EV system and will provide flexibility to consumers. Lithium and cobalt-based batteries will be used for storing electricity. An engineering mechanism is also employed for smooth functioning of the EV system. However, the EV system has become the best option for supplying electricity due to the presence of various flexibilities, such as the best transmission capacity, high flexibility, and availability of charging stations. Since EV is not only used for the energy source but also as electric load. In the field of electric vehicle, some improved response has been proposed, which can further enhance the feasibility and applicability of electric vehicles in a modern electric grid. Through the latest study, wireless power technology and an autonomous EV system have been introduced, which will be used for energy saving. Hence various smart cities are using such techniques and trying to find new solutions to address some of issues related to environmental, financial, and social [10].

In the future, the following recommendations need to be revised for improving and attaining feasible operation of the EV system:

1. Energy storage capacity needs to be revised and studied for further utilization.
2. Energy storage technology needs to be improved for charging of the EV system in a friendly manner.
3. Through grid connection, adverse effects of the EV charging system can be overcome.
4. Heating temperature of batteries needs to be reduced using air-cooling technology.
5. At a global level, acceptance of the EV module will take some time and its implementation can be improved by following the already employed EV countries.

REFERENCES

[1] Al-Hanahi, B., Ahmed, I., Habibi, D. and Masoum, M. A. S. "Charging infrastructure for commercial electric vehicles: challenges and future work." *IEEE Access*, vol. 9, pp. 121476–492, IEEE, 2021. https://doi.org/10.1109/ACCESS.2021.3108817.

[2] Vore, S. E., Kosowski, M., Reid, M. L., Wilkins, Z., Minicucci, J. and Bradley, T. H. "Measurement of medium-duty plug-in hybrid electric vehicle fuel economy sensitivity to ambient temperature." *IEEE Transactions on Transportation Electrification,* vol. 4, no. 1, pp. 184–189, IEEE, 2018. https://doi.org/10.1109/TTE.2017.2752131.

[3] Zhang, Z. J., Fang, W. and Ma, R. "Brief review of batteries for XEV applications." *eTransportation*, vol. 2, pp. 1–35, 2019. https://doi.org/10.1016/j.etran.2019.100032.

[4] Lukic, S. M., Cao, J., Bansal, R. C., Rodriguez, F. and Emadi, A. "Energy storage systems for automotive applications." *IEEE Transactions on Industrial Electronics*, vol. 55, no. 6, pp. 2258–67, IEEE, 2008. https://doi.org/10.1109/TIE.2008.918390.

[5] Simic, Z., Topic, D., Knezevic, G. and Pelin, D. "Battery energy storage technologies overview." *International Journal of Electrical and Computing Engineering Systems*, vol. 12, no. 1, pp. 53–65, 2021. http://dx.doi.org/10.32985/ijeces.12.1.6.

[6] Pareek, S., Sujil, A., Ratra, S. and Kumar, R. "Electric vehicle charging station challenges and opportunities: a future perspective." *International Conference on Emerging Trends in Communication, Control and Computing*, pp. 1–6, IEEE, 2020. https://doi.org/10.1109/ICONC345789.2020.9117473.

[7] Mojumder, Md. R. H., Antara, F. A., Hasanuzzaman, Md. Alamri B. and Alsharef, M. "Electric vehicle-to-grid (V2G) technologies: impact on the power grid and battery." *Sustainability*, vol. 14, no. 21, pp. 1–53, 2022. https://doi.org/10.3390/su142113856.

[8] Tiande, M., Li, Y., Lau, K., Poon, C. K., Wu, Y. and Luo, Y. "Trends and emerging technologies for the development of electric vehicle." *Energies*, vol. 15, no. 17, pp.1–34, 2022. https://doi.org/10.3390/en15176271.

[9] Nigam, A. and Sharma, K. K.. "Stability analysis of hybrid energy system using optimization technique." *International Conference on Computing, Communication, Security & Intelligent Systems*, pp. 1–6, IEEE, 2023. https://doi.org/10.1109/IC3SIS54991.2022.9885277.

[10] Hossain, M. S., Kumar, L., Haj Assad M. El and Alayi, R. "Advancements and future prospects of electric vehicle technologies: a comprehensive review." *Advanced Control and Optimization for Complex Energy Systems*, pp. 1–21, 2022. https://doi.org/10.1155/2022/3304796.

Chapter 2

Proof of concept

An on-road recharging e-vehicle

Ching Yee Yong and Chuan Zheng Ling

School of Engineering and Technology, University of Technology Sarawak, No. 1, Jalan University, Sibu, Sarawak, Malaysia

2.1 ELECTRIC VEHICLE (EV)

Electric vehicles (EVs) run on electricity, mostly the battery capacity of the electric vehicle is in the 130 km to 170 km distance range. Some luxury types of EVs come with a battery capacity that can withstand vehicles running up to 400 km. The demand for the EV has increased year by year and, undoubtedly, EVs will be the future of the automotive industry.

The advantages of using EVs were being cheaper to run and maintain, better for the environment, and most importantly, it helped to prevent the depletion of crude oil. Due to the need for crude oil in the market, the price of crude oil was expensive compared to electrical energy since crude oil was a kind of non-renewable energy that comes with a limited amount and might run out in the future. According to the statement of the organization of petroleum exporting countries, there were only nearly 1.5 trillion barrels of crude oil reserves left in the world [1], so the EV will be the best choice for future consideration. EVs come with zero exhaust emissions since no petrol combustion is performed in the engine for the car to move, thus reducing air pollution, which is better for the environment.

Everything comes with pros and cons. The disadvantages of the EV were the charging woes, its overloaded batteries, and the weight and price of the batteries [2]. We can see that most of the disadvantages come from the batteries of these EVs. Fuel-based vehicles require gas stations for the vehicle to 'charge' up, for sure EVs also require an electric charging station. So, to increase the possibility for the EV to travel a longer distance, the network of such stations has to be located strategically. Not only that, the recharging times for the EV normally take up to 3 hours in basic. These two were the charging woes that come with EVs. Other than that, the battery overloaded problems also bring some troubles for EV owners. Those accessories in the EV, such as radio, car air-conditioners, power windows, GPS, car security cameras, and some other kinds of accessories, shorten the time that electric power from batteries is used up and followed by the charging woes again [3]. The average lifetime for the batteries of EV is only around

DOI: 10.1201/9781003502470-2

4 years, and the cost of the batteries is much more expensive than the fuel-based vehicle's battery [4]. The batteries of the EV are costing thousands of dollars each. The weight of the EV's battery pack is up to approximately 450 kg, which becomes a disadvantage of the EV because the weight of the vehicle increases the pressure of the battery, which may result in the batteries draining up faster.

To overcome these problems that come from the disadvantages of the EV, a solution was proposed, which was to design a prototype of an EV that can move with a battery that can be charged while it is providing power for the vehicle to move. This system uses wireless power transfer (WPT) technology. In this project, the technology of WPT will be used for charging the battery of the vehicle while the vehicle is moving, so the battery of the vehicle can be used for a longer period and this can ease the problem of waiting for the long charging time of the battery. The method of WPT used in this project will be the inductive resonant coupling method because this kind of method can work in a medium range with higher frequency compared to normal magnetic inductive coupling. In simple words, the power will be transmitted from the transmitter installed on the road surface to the receiver, which is installed at the bottom of the EV prototype. By using this system, the problems of the typical EV might be greatly reduced, but we have to face the problem of whether the amount of power supplied to the EV is enough for the vehicle to operate normally or not. Table 2.1 shows the comparison between the existing and the proposed system. The proposed system provides a longer discharging period with a lower discharging rate. The novelty of this system is the EV prototype can charge while it is on the road without stopping and spending time at the charging bay.

Table 2.1 Comparison of different aspects between the existing system and the proposed system

Features	*Existing system*	*Proposed system*
Power transfer method	Used wireless transfer technology to charge the batteries and provide power to the electric motor	Used wireless transfer technology to provide power directly to the electric motor
Period for the battery to discharge	Standard	Longer
Discharge rate	Standard	Lower
Time used for charging	Minimum of 3 hours to fully charge the batteries	No need to spend time charging
Charging condition	Stationary without moving	Able to charge on moving

2.2 WIRELESS POWER TRANSFER (WPT) TECHNOLOGY

Wireless power transfer means that the power can be transferred, which will not involve any kind of cable during the transmission of the power. The power must be delivered from the power source to the desired load without any cable connected. Wireless power transfer techniques fall into two categories, which are radiative and non-radiative. The power is transmitted by magnetic fields using inductive coupling between coils of wire, and this kind of method is known as the non-radiative technique. In radiative techniques, it was used in the transmission for a wide range or longer distances. Two basic components must be involved in the system to allow the power to transfer wirelessly. These two components were the transmitter and receiver. There will be a short range of air gaps between the transmitter coil and the receiver coil. The types of non-radiative wireless power transfer techniques involved inductive coupling and resonant inductive coupling. The types of radiative wireless power transfer involved microwaves [5].

2.2.1 Comparison between different wireless power transfer methods for the E-vehicle prototype

Table 2.2 shows the comparison between different kinds of wireless power transfer methods based on the requirements of the EV prototype. The range of transmission, transfer frequency range, efficiency, antenna device, advantages, and disadvantages will all be discussed by using the table shown at the bottom.

2.2.2 Pros and cons of using wireless power transfer

By using wireless power transfer, the current system using a high-power transmission line can be eliminated, since transferring the power by using cable is very costly. The transmission system all over the world will become cheaper if wireless power transfer is further implemented. The large-scale reduction of power tariff can be easily noticeable and the loss of transmission will be lowered. Other than that, wireless power transfer techniques can increase safety purposes. The chances of electric shocks or short circuits will decrease, and humans and animals will be much safer since electric shocks by accidentally touching the cable will not exist anymore. Minimization of power failure will become achievable, which is caused by the short circuit and faults. The transmission system will become more efficient and environmentally friendly. Natural disasters such as floods, earthquakes, landslides, or tornados would not be the cause of power cuts and system damage [7].

The main concern about wireless power transfer is its biological effect and its initial cost for practical installation. The wireless power transfer using the microwave will bring health issues to humans due to the high

Table 2.2 Comparison between different wireless power transfer methods

Technology	*Range*	*Frequency*	*Efficiency*	*Antenna devices*	*Advantages*	*Disadvantages*
Induction coupling	Short	kHz	High	Wire coils	High efficiency	Short range
Resonant inductive coupling	Medium	kHz–MHz	Medium	Wire coils	Medium efficiency in short-range	Complex implementation
Microwaves	Long	GHz	Low	Rectenna	Long range	Radiative

Source: [6].

Table 2.3 Advantages and disadvantages of using wireless power transfer

Advantages	*Disadvantages*
(a) Lower cost (b) Reduction of power tariff can be easily noticeable (c) Safer (d) Minimization of power failure	(a) High initial cost (b) Interface of the microwaves with the present wireless communication system

Source: [7].

microwave radiation being high. Other than that, another potential disadvantage of wireless power transfer is the frequency used for wireless power transmission may interface with the present wireless communication system. Table 2.3 shows a summary of the advantages and disadvantages of using wireless power transfer [7].

2.2.3 Oscillators

An oscillator must be included in the circuit of the wireless power transfer system due to there being only a low frequency, which is 50 Hz on the present power supply and this was not sufficient to use in the wireless power transfer system. Due to that, an oscillator is used to increase the frequency up to kHz so that the power can be wireless transferred efficiently. In this section, two types of oscillators will be further discussed, which are the Colpitts oscillator and the Hartley oscillator.

2.2.3.1 Colpitts oscillator

Colpitts oscillator was invented by an American scientist named Edwin Colpitts in the year 1918. There are many types of Colpitts oscillators such as valves Colpitts oscillators, transistors-based, field-effect transistors (FETs)-based, or op-amp-based Colpitts oscillators. Colpitts oscillator is an electronic oscillator, which uses an inductor and capacitors to form an *LC* (inductor and capacitor) oscillator circuit [8].

The resistors R_1 and R_2 act as a voltage divider biasing to the transistor. The resistor R_4 is used to limit the collector current of the transistor. Other than that, there were five capacitors used in the circuit which were used for different purposes. Capacitor C_{in} acts as the input DC decoupling capacitor, while C_{out} is used as the output decoupling capacitor. An emitter by-pass capacitor is needed to bypass the amplified AC signals from dropping across R_e to prevent the DC biasing conditions of the transistor altered by the amplified AC signal and causing the transistor to become a reduced gain. The emitter by-pass capacitor was marked as C_e. Colpitts oscillators have an operating frequency range between 20 kHz to 300 MHz. The capacitive voltage divider setup in the tank circuit works as the feedback source and is thus able to bring better frequency stability to the oscillator circuit. The frequency is determined by the tank circuit and is varied by varying the values of the variable inductor L_1 and keeping the values of capacitors C_1 and C_2 [9,10].

The tank circuit is not only formed by the two capacitors, an inductor will be connected in parallel to the series capacitor. Capacitors C_1 and C_2 are the capacitors used in the tank circuit, and inductor L_1 is the inductor used for the tank circuit. Feedback to the base of the transistor is taken from the junction of capacitor C_2 and inductor L_1 in the tank circuit [11,12]. In the tank circuit, C_1 and C_2 start charging when the power supply is switched on. The capacitors will start to discharge through L_1 when they are fully charged. The electrostatic energy stored in the capacitors will be transferred as a magnetic flux to the inductor during the time that the capacitors are fully discharged. After that, the inductor starts discharging and the capacitors get charged again. The basis of oscillation can be understood as the transfer of energy back and forth between capacitors and inductors. The frequency of the Colpitts oscillator can be calculated by using the formula below.

$$f_r = \frac{1}{2\pi\sqrt{LC_T}} \tag{2.1}$$

$$C_T = \frac{C_1 \times C_2}{C_1 + C_2} \tag{2.2}$$

2.2.3.2 *Hartley oscillator*

The Hartley oscillator was invented by Ralph Hartley. Ralph Hartley was an American engineer, and he made this invention during the time that he was working for the Western Electric Company. The oscillation frequency is determined by the tank circuit, which comprises two series inductors in parallel with a capacitor in the Hartley oscillator [13]. The operating frequency range for the Hartley oscillator is between 20 kHz to 30 MHz. The Hartley oscillator can be operated at frequencies lower than 20 kHz, but the inductor value needs to be high, and it has a practical limit so that it is most suitable to apply at the frequency above 20 kHz.

The resistors R_1 and R_2 act as a role that gives a potential divider bias to the transistor Q_1. The resistor R_e, which connected to the emitter pin of the transistor, was an emitter resistor, and it was used to maintain the thermal stability of the transistor to prevent the transistor from overheating and burning. The capacitor C_e is an emitter bypass capacitor which does the job of bypassing the amplified AC signals [14]. Just like the Colpitts oscillator, an emitter bypass capacitor is needed to bypass the amplified AC signals from dropping across R_e to prevent the DC biasing conditions of the transistor altered by the amplified AC signal and causing the transistor to become a reduced gain. The other capacitors such as C_{in} in the circuit are used as the input DC decoupling capacitor, while C_{out} is used as the output decoupling capacitor.

When the circuit is switched on, the capacitor will start to charge. The capacitor will start to discharge through the inductor L_1 when it is fully charged. A series of damped oscillations in the tank circuit will be created through the process of charging and discharging the capacitor and inductor. The oscillation produced in the tank circuit will be fed back to the base pin of the transistor Q_1 in the amplified form across the collector and emitter of the transistor. The output voltage of the transistor Q_1 will be in the same phase as the voltage flow through the inductor L_1 [14]. The voltage across L_2 will be 180° out of phase to the voltage flow through L_1 and the voltage is fed back to the base pin of the transistor Q_1. So, the total phase difference between input and output voltage is 360° since the transistor will produce another 180° phase difference by itself. The formula used to calculate the frequency for the Hartley oscillator will be shown at the bottom.

$$f_r = \frac{1}{2\pi\sqrt{CL_T}} \tag{2.3}$$

$$L_T = L_1 + L_1 \tag{2.4}$$

C is the capacitor C_1, which is used in the tank circuit of the Hartley oscillator.

2.3 HALF-WAVE RECTIFIER

Half-wave rectifier is a type of rectifier, which only allows one half-cycle of the AC waveform to pass, and another half will be blocked and unable to pass through the diode. A half-wave rectifier was used to convert the AC power supply into DC form by using a single diode to use the desired load, a diode was connected to the input source. During the positive half-cycle, the diode is forward-biased, and the current is allowed to flow toward the load. During the negative half-cycle, the diode acts as reverse bias, and the current is blocked and unable to flow toward the load. The kind of half-wave rectifier can be considered the simplest kind of rectifier.

2.4 COIL

The design of the coil structure is one of the key points that needs to be considered for designing a wireless power system for an E-vehicle. The characteristics that indicated a good coil involved a high level of freedom, moderate coupling coefficient, less leakage, no heating of the coil, and a moderate distance between transmitting and receiving coils [15]. The coil used for wireless power transfer must be in one set, which consists of two different types of coils that are used as a transmitter coil and a receiver coil based on the requirement of the output load.

The coil construction will be based on several requirements to produce a high-efficiency wireless power transfer coil. The requirements that have to be met to construct a coil include the working frequency and transferring power, mechanical dimension of the coil, durable construction, and electromagnetic shielding [16], the yellow part that is the coil represents the core of the ferrite material.

2.5 SUMMARY

The literature review and case study related to wireless power transfer are done. Since there was several cm of separation between the transmitter coil and receiver coil in this project, the method that was decided to use for the E-vehicle prototype is resonant inductive coupling. The reason for choosing this method is that the resonant inductive coupling is available for longer distances in comparison to the inductive coupling. The mutual inductance of the coil is directly proportional to the efficiency of power transfer, so the ideal coil design for the transmitter system will be rectangular due to its higher mutual inductance compared to the circular coil. The oscillator that has been chosen for use in this project will be the Colpitts oscillator. The buck converter is selected as the DC-to-DC converter used on the receiving side to step up the current for the output load [17].

2.6 PROTOTYPING THE PROPOSED EV

2.6.1 Prototype overview

By referring to Figure 2.1, the AC power supply will be taken from the wall socket plug and will be converted into a DC power supply through an AC-to-DC converter. The supply must be converted to DC form because all the components used in the circuit are DC-rated, thus the power supply must be converted into DC. The DC power supply will then flow into the Colpitts oscillator to generate a high-frequency AC supply, which comes with a few hundred kilo Hertz. The transmitter coil and receiver coil do the job of transferring and receiving the power supply wirelessly. After the receiver coil receives the power supply, the power supply will flow to a half-wave

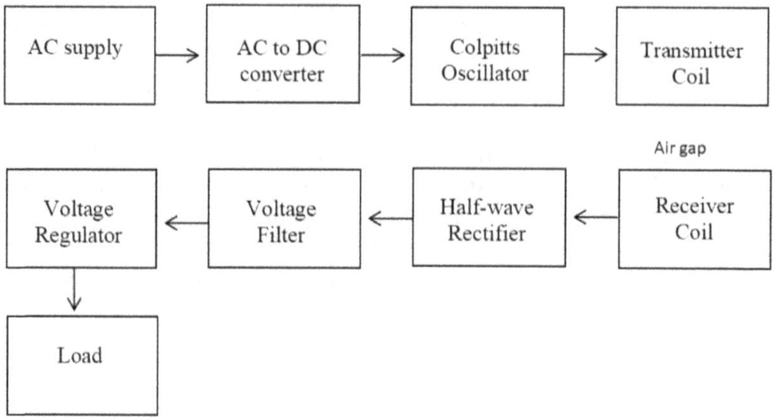

Figure 2.1 Block diagram of the wireless power transfer system.

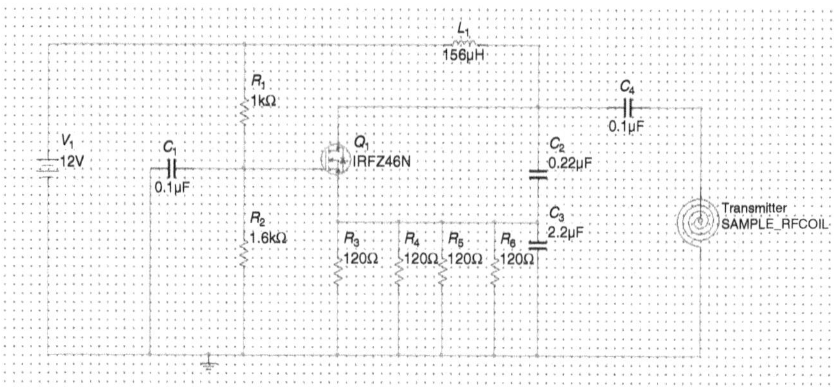

Figure 2.2 Schematic diagram of the transmitter circuit.

rectifier to turn the AC power supply to the DC supply for the desired load requirement. Before the power flows to the load, it will go through a voltage regulator to regulate the voltage, and also it will pass through a voltage filter to smoothen the induced voltage.

The transmitter circuit changes the AC into DC through an AC-to-DC converter. The DC power switches to high-frequency AC power after going through the oscillator circuit to transmit efficiently to the receiver part of the circuit. Figure 2.2 shows the schematic diagram of the transmitter circuit simulated from NI MULTISIM.

The receiver circuit receives the power transmitted by the transmitter coil. The power received was in AC form, thus the power supply was changed into DC by using a diode to perform half-wave rectification. The current is regulated by using an LM317 variable voltage regulator to regulate the voltage to the desired output voltage. Figure 2.3 shows the schematic diagram of the receiver circuit.

2.6.2 Function of each component in the transmitter circuit

By referring to Figure 2.3, the resistors R_1 and R_2 act as voltage dividers for the Q_1 transistor IRFZ46N. The value of resistor R_1 must be less than the value of resistor R_2 so that less current is passing through the base pin of the transistor. The purpose of parallel resistors, which are R_3, R_4, R_5, and R_6 in the circuit is to make the resistance at the emitter as small as possible so that the output voltage can be increased. These parallel resistors also function as heat dissipates to minimize the heat produced due to the high frequency of the transistor.

For the Colpitts oscillator, the tank circuit consists of two capacitors in series and an inductor connected in parallel to the serial combination. The inductor L_1 and the capacitors C_2 and C_3 were the Colpitts oscillator tank circuit. The main factor that determines the frequency of our voltage was the value used for the inductor and both capacitors. The inductor value can be calculated by using the following formula shown below:

$$C_T = \frac{C_4 \times C_5}{C_4 + C_5} \tag{2.5}$$

$$C_T = \frac{22\eta \times 220\eta}{22\eta + 220\eta} \tag{2.6}$$

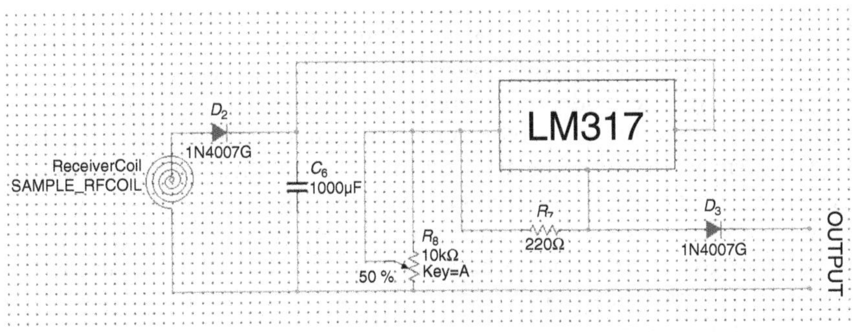

Figure 2.3 Schematic diagram of the receiver circuit.

$$C_T = 20\,\eta F \tag{2.7}$$

By setting the resonant frequency to 90 kHz as the desired output resonant frequency, and using the resultant capacitor value calculated above, the value of the inductor can be obtained through the calculation shown below.

$$f_r = \frac{1}{2\pi\sqrt{LC_T}} \tag{2.8}$$

$$90k = \frac{1}{2\pi\sqrt{L(20n)}} \tag{2.9}$$

$$L = 156.36\ \mu\text{H} \tag{2.10}$$

By rearranging the equation, the inductor value obtained through the calculation was approximately 156.36 uH. Since there was not the same value of the inductor as the calculated value in the market, the nearest value to the calculated value of the inductor will be chosen. After being surveyed, the inductor with the value of 156 uH was chosen.

Capacitors C_1 and C_4 act as bypass capacitors and coupling capacitors, respectively. The bypass capacitor C_4 was used to protect the power metal-oxide-semiconductor field-effect transistor (MOSFET) and speed up the output process by letting most of the gate current flow to the ground. The coupling capacitor C_1 was used to block the DC signal from the voltage source and produce a pure AC signal to the transmitter. Lastly, the transmitter coil was used to transfer the AC signal produced by the Colpitts oscillator to the receiver part of the circuit.

2.6.3 Function of each component in the receiver circuit

By referring to Figure 2.3, the receiver coil was used to receive the AC signal, which was transferred by the transmitter coil and induced the AC signal into the circuit. The diode in the circuit acts as forward bias at the positive half-cycle and allows the current to flow. The diode will act as reverse bias during the negative half-cycle and prevent the current from flowing through it, thus we can get the DC at the output. The capacitor C_5 acts as a voltage filter to produce a purer DC signal. The variable voltage regulator LM317 is used to regulate the output voltage to a maximum of 4.5 V to run the DC motor and charge the backup battery. The variable resistor R_7 is used to adjust the output voltage.

2.6.4 Summary of the section

In this section, the circuit design of this project was mentioned and explained. Colpitts oscillator had been chosen as the oscillator used for the circuit. The value of the capacitor is 22 and 220 nF, respectively, and the value of the inductor is 156 uH to produce a high oscillating frequency up to 90 kHz. A capacitor will be used in the receiver circuit so that the output DC voltage from the half-wave rectifier diode can be smoothed and delivered to the load. A voltage regulator will also be inserted in the receiver circuit to regulate the output voltage for the requirements of the load.

2.7 RESULT AND ANALYSIS

In this section, the design of the hardware will be proposed. Other than that, the results of stimulation, which is from NI MULTISIM and also the experimental results will be collected and shown in this section.

Figure 2.4 shows the proposed hardware design of the E-vehicle prototype. The transmitter coil will transfer the power to the receiver coil and the power flow to the receiver circuit. It will come out with DC power from the receiver circuit and provide power to charge the battery while moving. The

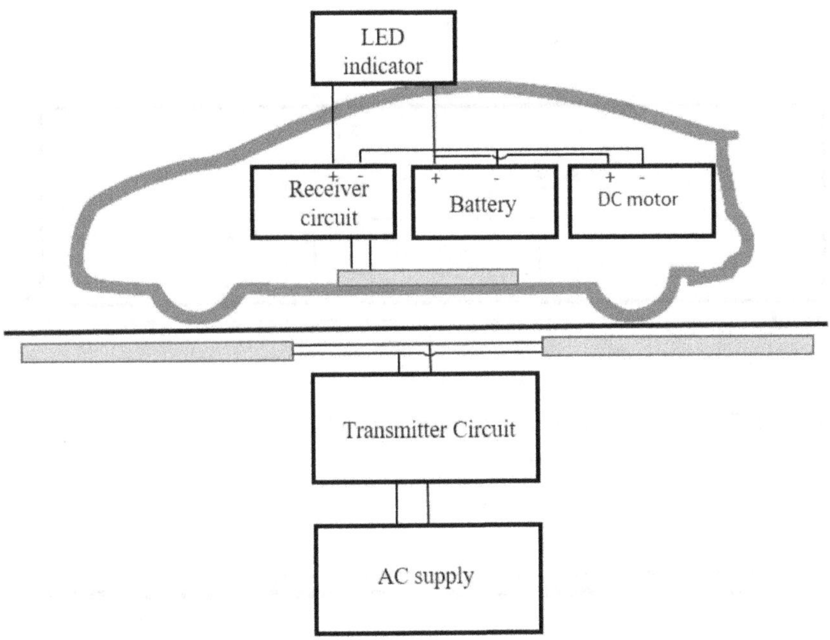

Figure 2.4 Proposed hardware design.

DC motor will start to work after receiving the power from a battery and turning the wheel of the vehicle.

2.7.1 Proposed hardware design of transmitter coil

Figures 2.5 and 2.6 show the top view and the side view of the proposed transmitter coil, respectively. A 25 cm by 40 cm transmitter coil was designed for the transmitter circuit. The coil will be mounted on the surface of a 1.5-meter-long wood. The transmitter circuit will be placed at the side of the transmitter coil.

2.7.2 Proposed hardware design of receiver coil

Figure 2.7 shows the proposed hardware design of the receiver circuit. The receiver coil is attached to a piece of the ferrite coil. The receiver coil and the receiver circuit are both installed in the chassis of the E-vehicle prototype. The receiver circuit will transfer the DC power to the DC motor.

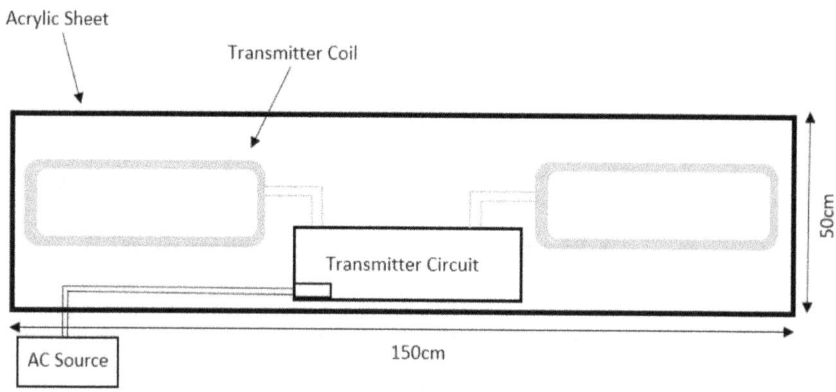

Figure 2.5 Top view of the proposed transmitter coil.

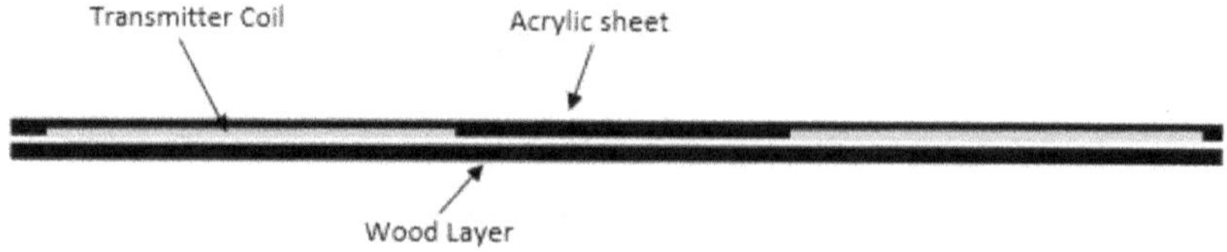

Figure 2.6 Side view of the proposed transmitter coil.

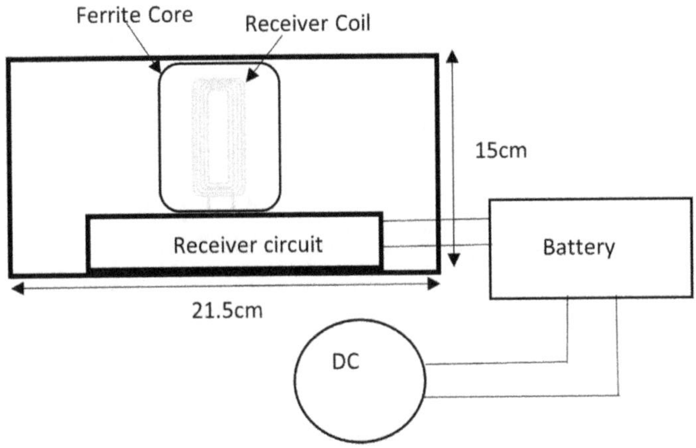

Figure 2.7 Proposed hardware design of receiver circuit.

2.7.3 NI MULTISIM simulation result

The simulation is done by using the NI MULTISIM and the frequency after the Colpitts oscillator is around 95 kHz, which is slightly different from the calculation results. This might be due to the tolerance in the components used in the circuit design. The output AC voltage at the transmitter coil has a peak value of around 25 VAC. The output waveform of the Colpitts oscillator is shown in Figure 2.8 and the high-frequency waveform is counted as successfully generated.

Figure 2.9 shows the output DC waveform from the receiver circuit after the filter capacitor. Based on the simulation result in NI MULTISIM, the voltage at the output of the receiver is around 20.5 VDC. Other than that, the waveform shown in the NI MULTISIM can be counted as a perfect DC waveform after being filtered by the capacitor.

Figure 2.10 shows the constructed hardware. The EV prototype was on the top of the lane, which was installed with a transmitter coil. The LED lights up when the EV prototype receives the power from the transmitter coil to charge the battery.

2.8 THE PROPOSED EV STRENGTHS AND LIMITATIONS

This project mainly is to design a low-cost wireless power transfer system and also to distribute the concept of the E-vehicle which can run by using wireless power transfer. To achieve wireless power transfer, the frequency of the AC voltage has to be extremely high, so the 50 Hz AC source from the wall plug will be converted into a high-frequency AC source by using

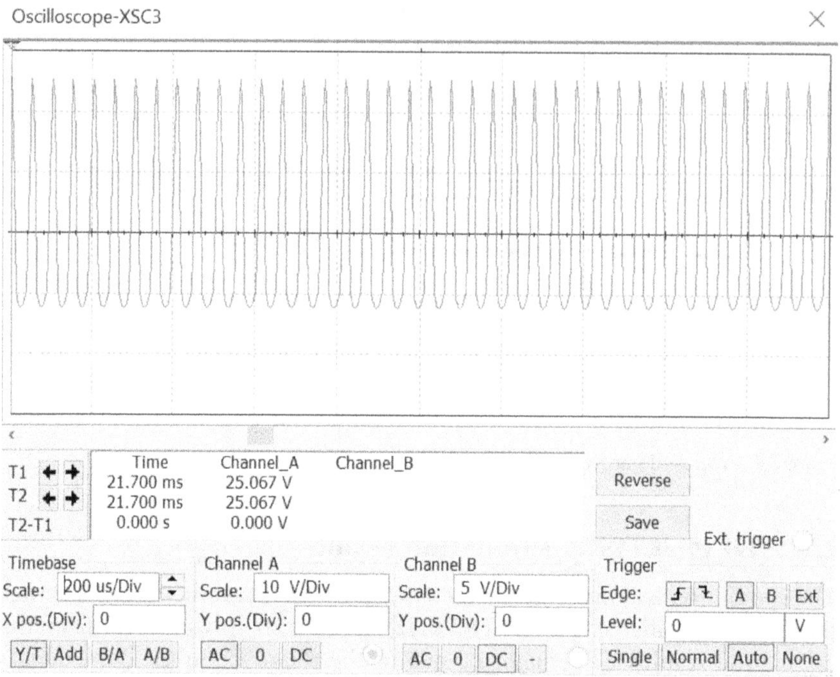

Figure 2.8 Transmitter circuit output AC waveform.

the Colpitts oscillator. The frequency used for wireless power transfer was determined by the tank circuit of the Colpitts oscillator. Based on the study, some cases use the same transmitter circuit to generate 500 kHz high frequency for the power to transfer wirelessly. After trial and error, this project had been decided to use 90 kHz for the wireless power transfer. The reason for choosing 90 kHz is due to at this frequency range, it can perform much better than other higher frequencies, and it can deliver more current to the receiver part.

According to the summary table of the results (refer to Table 2.4), the current delivered to the receiver side was 19.63 mA, which is much higher than the current delivered at other higher frequencies. The reason the current is lowered when using a higher frequency is due to the heat generated at a higher frequency will link to the power losses, thus affecting the output current obtained at the receiver side. The distance for the power to transfer wirelessly will not be affected when choosing 90 kHz, even if the frequency was much lowered.

Other than that, the transistor used at the transmitter side has also been selected after proper consideration and trial and error to get the best

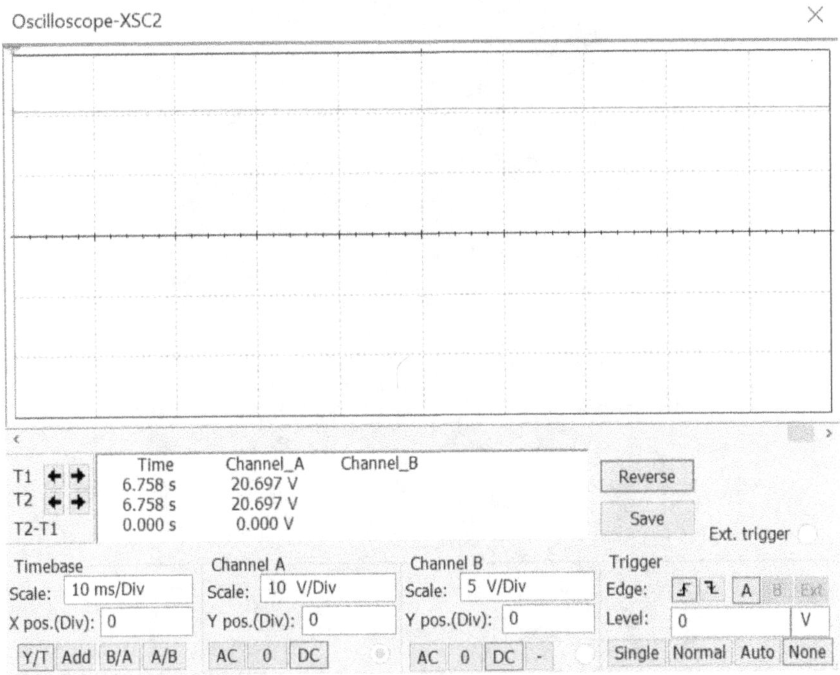

Figure 2.9 Receiver circuit output DC waveform after filter capacitor.

Table 2.4 Summary output from the NI MULTISIM

Output	*Voltage (V)*	*Current (A)*	*Frequency (Hz)*	*Power (W)*
Socket outlet	240 VAC		50	
Adapter	12 VDC	2		24
Colpitts oscillator	25.07 VAC		90 k	
Receiver coil	25.07 VAC		90 k	
Half-wave rectifier	20.7 VDC			

transistor selection for this project. In this project, the transistor used was an IRFZ46N power MOSFET transistor, the reason for choosing this transistor is due to it having a fast switching capability compared to the other types of NPN transistors. Using this transistor, the circuit can generate more high-frequency current at the transmitter side. Due to the fast switching, the transistor will get heat easily, so it will assemble on a heat sink to dissipate the heat generated. Not only that, the heat will also be generated at the source pin of the transistor. To ease this problem, a high-power resistor will

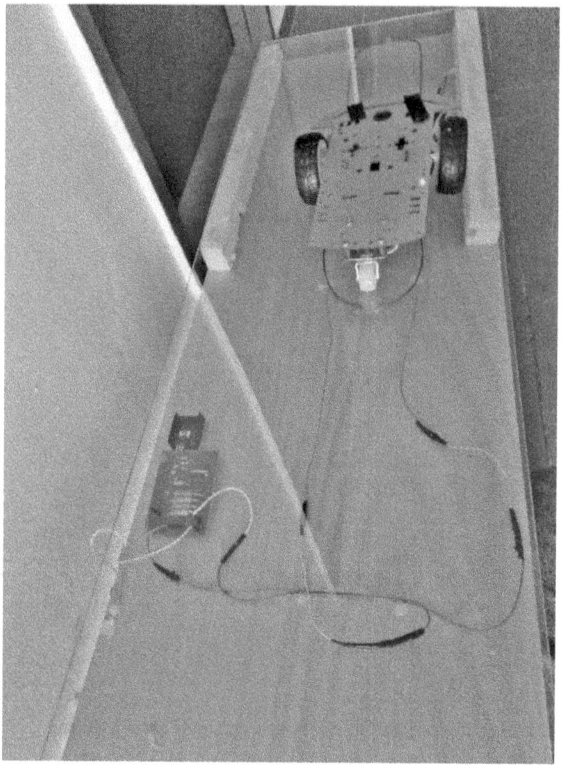

Figure 2.10 Overview of prototype.

be connected in parallel to dissipate the heat generated. The power rating of the resistor will be suggested as 2 watts, if not the resistor will also get heated and burnt.

According to the summary table of the result, the voltage generated at the receiver coil is higher than the voltage generated at the transmitter coil due to the number of turns of the coil being slightly higher than the transmitter coil at the same diameter of the coil. Lastly, the method for rectification at the receiver side was using half-wave rectification. One of the reasons for using half-wave rectification is due to it only requiring one diode to operate, instead of four diodes. The other reason is that the efficiency of using a bridge rectifier is not as ideal as using a half-wave rectifier in this project and the reason for this might be due to the power quality problem, which it can operate normally only at positive half-cycle, but it cannot work nicely at negative cycle.

2.9 CONCLUSION AND FUTURE WORK

To conclude this project, the objective can be considered as achieved successfully. There are several types of wireless power transfer methods but in this project, inductive coupling methods were chosen since this project was mainly to introduce the concept of the E-vehicle prototype and design a low-cost wireless power transfer system, in this case, inductive coupling is a good selection for it.

At the transmitter part, the selection of the oscillator type is one of the important factors that need to be considered so that the 50 Hz AC can be switched into the desired high-frequency AC for the transmission purpose. In this project, the Colpitts oscillator had been chosen as the oscillating circuit as it was able to provide good stability at high frequency and the operating frequency range was wide enough for this project to generate the desired high frequency. Other than that, the transistor selection for the oscillator is also very important to perform switching and oscillate the input signal continuously. The transistor used in this project was IRFZ46N. It was a power MOSFET transistor, and it performed better than other NPN transistors such as 2N2222. Both two transistors have been tested in hardware and IRFZ46N got better results compared to the 2N2222.

The design of the coil was also one of the key factors in the world of wireless power transfer. The length of the coil was significant for the coil design. The coil used for this project was designed to overlap between each turn of the coil instead of the spiral coil. This kind of design was chosen because the spiral design coil only provides high performance at the center of the coil but at the side, it was unable to provide the expected efficiency. The number of turns for the transmitter coil was five turns but with a bigger diameter, and the number of turns for the receiver coil was 40 turns with a smaller diameter. This kind of design made the receiver coil induce more voltage from the transmitter coil by the proof of results in the following section. The voltage at the transmitter coil was 28.3 V and the receiver voltage was 37.6 V, the voltage at the receiver was higher than the voltage at the transmitter, which means the coil design worked successfully.

The voltage is then converted into DC to provide power for the desired output load. Half-wave rectification was done with a filter capacitor in this project to rectify the AC to DC. A variable voltage regulator is also added to the receiver side to regulate the voltage to the desired output voltage. The reason for using a variable voltage regulator is due to the minimum voltage regulator in the market being LM7805, and the regulated voltage was still considered slightly high for the output load in this project.

The recommendations can be used to increase the efficiency of the wireless power transfer system, and the concept of this project included adding receiver ICs at the receiver side. The receiver IC can provide better AC to

DC conversion, so we might get better output by using it. Other than that, the current at the receiver part can be increased to increase the period for the battery to discharge fully. On the transmitter side, high-frequency and high-power ICs can be used for high-frequency switching purposes instead of the transistor since the ICs might provide good stability on switching high frequency compared to the transistor.

In conclusion, the outcome of this proposed prototype can undergo a series of rigorous testing and validation in terms of safety, technical feasibility, and performance evaluation to meet the industry standards for commercialization. Collaboration between the university and industry can be done to refine the design to address the shortcomings. Additionally, the industry can focus on optimizing production by ensuring reliability, scalability, and cost-effectiveness.

REFERENCES

1. Ladopoulos, E. G. "Real-time expert seismology by non-linear oil reserves expert system." *Journal of Petroleum and Gas Engineering* 4, no. 2 (2013): 28–34.
2. Zhan, Weipeng, Zhenpo Wang, Lei Zhang, Peng Liu, Dingsong Cui, and David G. Dorrell. "A review of siting, sizing, optimal scheduling, and cost-benefit analysis for battery swapping stations." *Energy* (2022): 124723. https://doi.org/10.1016/j.energy.2022.124723
3. Chavan, Santosh, B. Venkateswarlu, R. Prabakaran, Mohammad Salman, Sang Woo Joo, Gyu Sang Choi, and Sung Chul Kim. "Thermal runaway and mitigation strategies for electric vehicle lithium-ion batteries using battery cooling approach: A review of the current status and challenges." *Journal of Energy Storage* 72 (2023): 108569. https://doi.org/10.1016/j.est.2023.108569
4. Ajanovic, A., and R. Haas. "Economic and environmental prospects for battery electric-and fuel cell vehicles: a review." *Fuel cells* 19, no. 5 (2019): 515–529. https://doi.org/10.1002/fuce.201800171
5. Triviño, Alicia, José M. González-González, and José A. Aguado. "Wireless power transfer technologies applied to electric vehicles: a review." *Energies* 14, no. 6 (2021): 1547. https://doi.org/10.3390/en14061547
6. Vinge, Rikard. "Wireless energy transfer by resonant inductive coupling." (2015). https://odr.chalmers.se/server/api/core/bitstreams/4ff48461-c8c9-4004-acec-3c1ac3dc2ce7/content
7. El Rayes, Mohamed M., Gihan Nagib, and W. G. Ali Abdelaal. "A review on wireless power transfer." *International Journal of Engineering Trends and Technology (IJETT)* 40, no. 5 (2016): 272–280. https://doi.org/10.14445/22315381/IJETT-V40P244
8. Hart, Daniel W., and Daniel W. Hart. *Power electronics*. Vol. 166. New York: McGraw-Hill (2011). ISBN-13: 978-0073380674.
9. Zhang, Zhuhaobo, Fan Zhu, Dehong Xu, Philip T. Krein, and Hao Ma. "An integrated inductive power transfer system design with a variable inductor

for misalignment tolerance and battery charging applications." *IEEE Transactions on Power Electronics* 35, no. 11 (2020): 11544–11556. https://doi.org/10.1109/TPEL.2020.2987906
10. Galandanci, G. S. M., and K. O. Ewansiha. "Design and simulation of a 20 kHz to 50 kHz Variable Frequency Oscillator (VFO)." *International Journal of Research and Reviews in Applied Sciences (IJRRAS)* 16, no. 1 (2013): 24–38. https://doi.org/ 10.11591/ijeecs.v1.i2.pp294-299
11. Nyamieri, Dominic Ondieki, and Cliff Mosiori. "Design and implementation of FM transmitter for security and communication purpose at the coast General Hospital wards." *Available at SSRN 4537788* (2023). https://doi.org/10.2139/ssrn.4537788
12. Slusarczyk, Adrian L., Allen Lin, and Ron Weiss. "Foundations for the design and implementation of synthetic genetic circuits." *Nature Reviews Genetics* 13, no. 6 (2012): 406–420. https://doi.org/10.1038/nrg3227
13. Garinto, Dodi, Ary Syahriar, and Setiyo Budiyanto. "A novel op-amp based LC oscillator for wireless communications." In *IECON 2020 The 46th Annual Conference of the IEEE Industrial Electronics Society* (2020): 2243–2248. IEEE. https://doi.org/10.1109/IECON43393.2020.9255128
14. Malvino, Albert Paul, David J. Bates, and Patrick E. Hoppe. *Electronic principles*. Glencoe: McGraw-Hill (1993). ISBN 978-0-07-337388-1.
15. Ongayo, Daniel, and Moin Hanif. "Comparison of circular and rectangular coil transformer parameters for wireless power transfer based on finite element analysis." In *2015 IEEE 13th Brazilian Power Electronics Conference and 1st Southern Power Electronics Conference (COBEP/SPEC)* (2015): 1–6. IEEE. https://doi.org/10.1109/COBEP.2015.7420222
16. Choi, Su Y., Beom W. Gu, Seog Y. Jeong, and Chun T. Rim. "Advances in wireless power transfer systems for roadway-powered electric vehicles." *IEEE Journal of Emerging and Selected Topics in Power Electronics* 3, no. 1 (2014): 18–36. https://doi.org/10.1109/JESTPE.2014.2343674
17. Yong, Ching Yee, and Kok Fen Chen. "Wireless power transfer technology using resonant technique." *IOP Conference Series: Earth and Environmental Science* 268, no. 1 (2019): 012102. https://doi.org/10.1088/1755-1315/268/1/012102

Chapter 3

Cylindrical lithium-ion battery

Fundamentals, design, and manufacturing processes

Vishal Dabra,[1] *Amit Kumar Dubey,*[1] *Ajay Kumar,*[2] *and Parveen Kumar*[3]

[1]Mechanical Engineering Department, P.I.E.T, Samalkha, India
[2]Department of Mechanical Engineering, School of Engineering and Technology, JECRC University, Jaipur, India
[3]Department of Mechanical Engineering, Rawal Institute of Engineering and Technology, Faridabad, India

3.1 INTRODUCTION

The automotive sector is expanding globally and provides engineers with a common package. The use of lithium-ion batteries (LiBs) is increasing at the moment due to circumstances where there is a growing demand for electric vehicles (EVs) to reduce air pollution caused by burning petroleum products [1, 2]. LiB is a very crucial component of any type of EV because it supplies energy/current to the motor, which helps to accelerate the wheels. Due to its unique qualities such as its high energy density, high operating voltage, and extended lifespan, LiBs are becoming more and more common in EVs. Numerous research projects are underway to improve the functionality and architecture of LiBs [3–5].

Technology has advanced remarkably since the LiB was first introduced in 1991, making it the most efficient source of reusable batteries available now and in the near future. Even so, EVs are still in the early stages of development. The researchers think that raising energy density is not as important as enhancing LiB lifetime performance. The Li-ion cell unit has garnered attention from researchers as a high-energy battery [6].

Nissan and Sony engineers explained its creation and noted that because of the high cell voltage, only a small number of cells are required, improving battery management. Based on voltage monitoring, accurate battery state-of-charge (SOC) detection is feasible. According to reports, Nissan introduced the Ultra EV in 1999. The vehicle would reportedly have LiBs and have a 120-mile range between charges. Lithium-polymer batteries are predicted to have 300-mile ranges even in the far future [7]. This chapter describes the fundamental concepts, design, and manufacturing processes of cylindrical LiBs.

 DOI: 10.1201/9781003502470-3

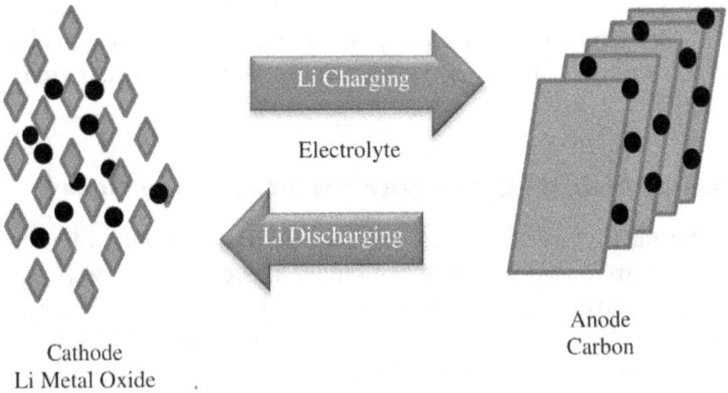

Figure 3.1 Schematic diagram of basic electrochemical reaction.

3.2 FUNDAMENTALS OF CYLINDRICAL LIB

In the study of the periodic table, lithium is the lightest of all the metals found on Earth. It shows excellent electrochemical properties, which make it more popular in rechargeable battery creation technology [8].

Two electrodes, the cathode and anode, are where electrochemical conversion takes place. The chemistry of the electrode determines the type of reaction. The energy content of the battery is dependent on the mass and volume of the lithium but the power of the battery is highly influenced by the area of the electrodes [9]. Basic electrochemical reaction is explained with the help a schematic diagram as shown in Figure 3.1.

LiB consists of multiple layers of different materials, which perform like electrodes (cathode and anode), separator and electrolyte. Instead of using metallic lithium, the LiB employs lithiated transition metal intercalation oxide for the cathode, lithiated carbon intercalation material for the anode, and a liquid organic solution or the solid polymer for the electrolyte.

During charging and discharging process, lithium ions oscillate between the cathode and anode in the electrolyte [10, 11]. During charging process movement of lithium ions are from cathode to anode and on the other hand during discharging process, lithium ions are discharged from the anode, move through the electrolyte, and are taken up by the cathode. It is expected that the improvement in LiB will lead to the manganese-based type due to their low cost, easily availability, and eco-friendliness nature.

LiB batteries tend to have lithium, cobalt, and nickel in their cathodes, graphite in their anodes. For other cell and pack components, aluminium and copper is preferred. Most commonly, NCM and LFP are used as a cathode in LiBs. NCM represents lithium nickel cobalt manganese oxide and LFP represents lithium iron phosphate. The instant revolution in battery technology, innovative chemistries can be anticipated in the future [12].

3.3 BASIC PRINCIPLE, THEORY, AND CALCULATIONS

Before starting to make LiB, first we require data related to battery technology which are based on basic principles. Basic electric law (Ohm's law) is used in the starting calculation of the battery pack. Generally, it provides the relationship between the three parameters in an electric framework, which are voltage, current, and resistance. [13].

In this chapter, voltage, current, and resistance are denoted by V, I, and R, respectively. Mathematically, Ohm's law is explained in equation (3.1).

$$V = I \times R \tag{3.1}$$

From equation (3.1), two parameters are found of the three parameters and we will be able to calculate the lost one. The equivalent circuit of Ohm's law in electrical engineering is shown in Figure 3.2. In battery calculation, several steps are involved as shown in Figure 3.3 [14].

3.3.1 Determining the required number of cells

The calculation of the number of cells required to build a battery pack is critical before beginning the fabrication of a Li-ion battery pack. The total number of cells required to match the fundamental voltage and current of the battery pack is the initial step in the battery calculation. The specified pack voltage is equal to and greater than the voltage required to run the electric drive (motor) within the framework voltage. Mathematically, how many cells are required in the battery pack is identified by dividing the V_p by the V_c as stated in equation (3.2).

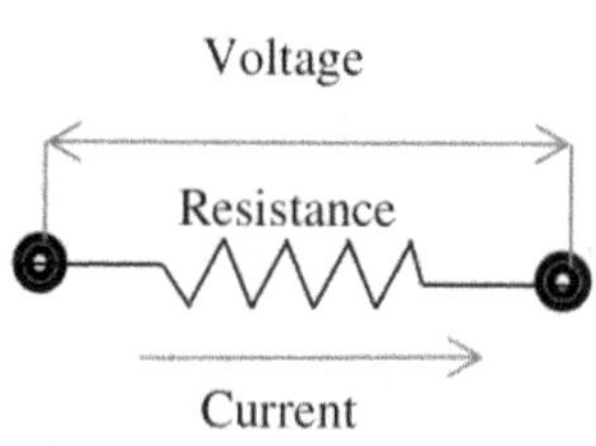

Figure 3.2 Electric circuit diagram of Ohm's law.

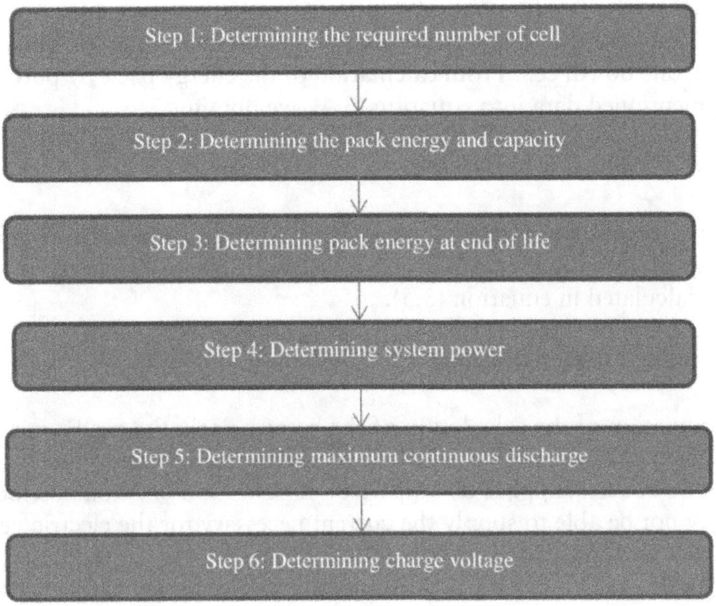

Figure 3.3 Flow diagram of battery calculation.

$$\text{Number of cells in series connection} = V_p/V_c \tag{3.2}$$

As we know, the most commonly available lithium cells in the market are NMC and LFP. For illustration, a battery pack of 120 V is desired and we employ the commonly available NMC 3.6 V lithium ion cell (cylindrical shape). By calculating the number of cells it is found that 120 V/3.6V = 33.3 cells. After rounding off to get the desired voltage, 34 cells are arranged in the battery pack. Additionally, to realize a 120 V pack, we employ the commonly available LFP 3.2 V lithium ion cell (cylindrical cell). By calculating the number of cells it was found that 120 V/3.2 V = 37.5 cells. After rounding off to get the desired voltage, 38 cells are arranged in the battery pack. In the NMC illustration, two modules of 17 cells each are used to manage 34 cells.

3.3.2 Determining the pack energy and capacity

The pack energy (E_p) is calculated by multiplying the pack voltage (V_p) with current (I_p). Mathematically it is stated in equation (3.3).

$$E_p = V_p \times I_p \tag{3.3}$$

The 20 kWh pack is assumed in this illustration and we employ 3.7 V nominal NMC cells. To obtain V_p of 334 V, 90 cells are connected in series. To get the desired energy, 60 Ah of cells are required and may be finished by utilizing one 60 Ah cell. From calculation of the energy pack by putting the above mentioned data into equation (3.4), we obtain

$$334 \text{ V} \times 60 \text{ Ah} = 20.040 \text{ kWh} \tag{3.4}$$

For the instance of an electric vehicle that is focusing on around 60 miles of running extent with a framework that is able to give 3 kWh/mile productivity is calculated in equation (3.5).

$$60 \text{ miles}/3 \text{ kWh/mile} = 20 \text{ kWh} \tag{3.5}$$

In continuation of the calculation of the battery pack, the prediction about the quantity of current required to run an electric motor is very much significant. A 60 Ah cell supplies 60 ampere of current for a one hour period. But this may not be able to supply the current necessary for the electric vehicle. However, this may lead us to talk about 'C-rates'.

Generally, a C-rate is provided by the cell manufacturer to provide details about the current requirement during cell charging and discharging processes. So for a better understanding, take a cell of 60 Ah with a 1C rate, which clearly explains that 60 ampere of current can be drawn for a one hour period. Instead of that, if same rated cell is operated at the 5C rate for a period of 10 seconds, which reflects that 300 ampere of current can be drawn for the 10 second period. Similarly, C-rate is equally significant for charging of the cell.

There are several formulas derived from Ohm's law to calculate the pack energy and voltage. As we know that Ohm's law gives $R = V/I$ and here the same formulas are used to calculate the current, pack energy and voltage as stated in equation (3.6), (3.7), and (3.8).

$$20 \text{ kWh}/334 \text{ V} = 59.9 \text{ Ah} \tag{3.6}$$

$$334 \text{ V} \times 59.9 \text{ Ah} = 20 \text{ kWh} \tag{3.7}$$

$$20 \text{ kWh}/59.9 \text{ Ah} = 334 \text{ V} \tag{3.8}$$

3.3.3 Determining pack energy at end of life

Regularly, we are able to expect that 100% of the battery is expended amid releasing but in real practice only 80–90% of the battery is used. In the given illustration, 20 kWh of energy is required from the battery pack. In the safe side of the battery pack, 80% of the complete energy is utilized during operation and the remaining 20% of the energy is unused in the battery pack.

The calculation of 80% utilization of the battery pack is illustrated with the help of equation (3.9).

$$20\ \text{kWh}/80\% = 25\ \text{kWh} \tag{3.9}$$

From equation (3.9), it is clearly observed that if 20 kWh of energy is required to run 60 miles but the battery pack of 25 kWh is used as per the 80% consumption condition of the battery pack. A common question asked to consumer is how much usable energy is required. Therefore by giving all of these issues our attention, a 25 kWh battery pack is supplied, which is larger than the actual energy needed. For calculating the capacity of the cells, the relationship of pack energy and pack voltage is stated in equation (3.10).

$$E_p/V_p = I_p \tag{3.10}$$

Therefore, by using equation (3.10) we find

$$20\ \text{kWh}/334\ \text{V} = 60\ \text{Ah cell} \tag{3.11}$$

In this illustration, 90 cells of 3.7 V are required to produce the 334 V battery pack at 60 Ah in capacity as calculated in equation (3.11). This is able to provide 25 kWh of total energy. The difference of total energy and useful energy is observable due to the fact that all LiBs available nowadays are not consumed 100%.

3.3.4 Determining system power

For calculating system power, Ohm's law and Joule's law are used in combined form. The formula used in the calculation of electric power is stated as in equation (3.12). It is calculated in watts.

$$P = I^2 \times R \tag{3.12}$$

where P represents the power (in watts), I represents the current (in amperes), and R represents the resistance (in Ohms).

For illustration, a cell of 7 mΩ of internal resistance is used to calculate the power is stated as in equation (3.13).

$$60\ \text{Ah} \times 7\ \text{m}\Omega = 420\ \text{W per cell} \tag{3.13}$$

By rearranging the equation (3.12) for calculating the electric power by using voltage and resistance is stated as in equation (3.14).

$$P = I^2 R = V\,(V/R) \tag{3.14}$$

And reverse of equation (3.14) may also be used to calculate I and V as stated in equations (3.15) and (3.16).

$$I = P/V \tag{3.15}$$

$$V = \sqrt{(P \times R)} \tag{3.16}$$

From the above stated equations, the value of voltage, current, resistance, and power may be calculated and explained with an illustration as follows in equation (3.17).

$$90 \text{ cells} \times 420 \text{ W} = 37800 \text{ W or } 37.8 \text{ kW} \tag{3.17}$$

As the power of the cell is mentioned on the cell and published in the data sheet by the cell manufacturer. If in any case the data is not given by the cell manufacturer, we can ask them to provide it.

The new term 'peak power' is also calculated to provide the power capabilities of the battery pack. This could be calculated in a simple and short period of 10 s, 5 s, 2 s, or 1 s. In general it is calculated in two parts. Firstly, the system resistance is calculated by dividing the change in the voltage by the change in current. The relationship among that is stated as in equation (3.18).

$$\text{Resistance} = \Delta V / \Delta I \tag{3.18}$$

The peak power (kW) is calculated by using above value of resistance as stated in equation (3.19). In this equation, peak power is calculated by dividing the square of the maximum open circuit voltage by four times the resistance.

$$\text{Peak power} = V^2 / 4R \tag{3.19}$$

3.3.5 Determining maximum continuous discharge current

In continuation of battery calculation, maximum continuous discharge is calculated by multiplying the number of cells in parallel (N_p), cell current (I_c), and the maximum C-rate (C_{max}) as stated in equations (3.20) and (3.21). Maximum continuous discharge current is denoted by I_{max}.

$$N_p \times I_c \times C_{max} = I_{max} \tag{3.20}$$

or

$$1 \times 90\text{A} \times 5\text{C} = 450\text{A} \tag{3.21}$$

3.3.6 Determining charge voltages

Lastly, in battery calculation, maximum charging voltage is calculated as stated in equation (3.22) as data provided by the cell manufacturer.

$$96 \text{ cells} \times 4.2 \text{ V} = 403 \; V_{max} \tag{3.22}$$

Similarly, the minimum charging voltage is calculated as stated in equation (3.23) as data provided by the cell manufacturer.

$$96 \text{ cells} \times 2.7 \text{ V} = 259 \; V_{min} \tag{3.23}$$

3.4 LIB HEATING AND COOLING REQUIREMENTS

The majority of LiBs operate at room temperature but a few operate at higher temperature. It is required to warm up initially and then cool down while in use. In some instances, the performance of batteries dropping off at low temperatures is undesirable, but this issue could be overwhelmed by heating up the battery [15]. The best range of temperature is between 15°C and 35°C for increased lifespan of LiBs [16]. The designer must be aware of the battery's capabilities when choosing it. The EV design process requires consideration of the temperature, heating, and cooling needs of LiBs.

3.5 BATTERY ENDING MEASURES

There are various characteristics of the ideal battery, which must be kept in mind. Battery terminals must offer low resistance along with tight connection in the electric system. During temperature change and vibration, terminal connection remains unaffected. Polarity should be kept in consideration while working with the battery [17]. In addition to proper working of the battery, BMS is connected to monitor battery performance. By this we are able to observe the voltage, current, and temperature of the battery. It also helps to detect the battery fault and state of charge.

3.6 MANUFACTURING PROCESSES AND NOMENCLATURE OF LI-ION CELL

There are various manufacturing processes for the lithium ion cell out of which designing the cell is a primary and important step. Basic requirements must be kept in mind during designing of cells as discussed below:

1. The electrode acts as the cathode and the anode must be positioned face to face.
2. Both act as electrically isolated to each other.

3. The capacity of both is the same and the distance between them isalso uniform.
4. Uniform electrical conductivity must be ensured throughout the cell.

Generally, the manufacturing process of the lithium ion cell consists of three basic processes, which are discussed below [18]:

- Production of electrode (cathode and anode).
- Assembly of the lithium ion cell.
- Cell forming, aging, and testing.

LiBs have a unique way of being named and the most common design is mentioned in equation (3.24).

$$N_1A_1A_2A_3N_2/N_3/N_4\text{-}N_5 \tag{3.24}$$

where

- N_1 stands for the number of cells joined in series and N_5 stands for the number of cells joined in parallel.
- A_1 stands for negative electrode phase. *I* stands for lithium-ion and L stands for lithium metal or alloy.
- A_2 stands for positive electrode phase. V, T, M, N, and C stands for vanadium, titanium, manganese, nickel and cobalt, respectively.
- A_3 stands for the shape of the cell. *R* stands for cylindrical cell and *P* stands for prismatic cell.
- N_2 stands for cylindrical cell diameter and for prismatic cell it represents thickness in mm.
- N_3 stands for prismatic cell width in mm.
- N_4 stands for overall height in mm.

Industry also uses one more nomenclature for representing lithium-ion cylindrical cells: D_1H_1

Here D_1 means the diameter of the cell (in mm) and H_1 means height of the cell (in mm) as shown in Figure 3.4.

The most common cylindrical shapes of lithium ion cells available in the market are listed in Table 3.1.

3.7 TECHNOLOGICAL CHALLENGES ASSOCIATED WITH MANUFACTURING PROCESSES OF CYLINDRICAL LIB

There are several technical challenges faced by lithium-ion battery manufacturers and some of them are discussed in this section. Establishing

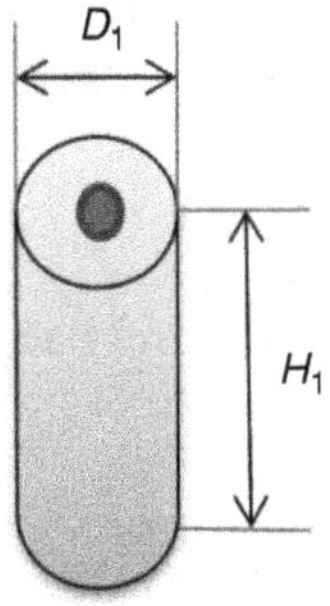

Figure 3.4 Cylindrical shape lithium ion cell.

Table 3.1 Description of cylindrical lithium ion cell

Cell description	*Diameter of the cell*	*Height of the cell*
18650	18	65
21700	21	70
26650	26	65

a production concept that is integrated, technologically balanced, and uses many interfaces and technologies presents a major challenge for cell manufacturers. The manufacturing of LiBs typically involves multiple production technologies in a wide range of expertise, unlike module and pack assembly. Cell production machines and plants are marketed by various companies for this reason. The popular cell production line is covered by only a few machineries and plant manufacturers with their own technology.

There are many companies that have expertise in a specific production line and provide different fabrication technologies for sub-processes in battery cell production. These companies are lacking in knowledge of specific production line and required specialist for establishing a specific production line.

The challenges of lithium-ion cell manufacturers are to find the right machineries and plant manufacturers for each process line. To ensure that the overall process is cost-effective, suppliers and manufacturers mustdevelop new machineries. This helps to simplify the production line working and process for a greater production rate. That is why iithium-ion cell manufacturers are transforming their manufacturing process into a mass production process by trying to adopt automation technology. The superiority of the product is significantly different because of the existing production structures.

Lithium-ion cell destruction has three shortcomings [1]:

1. The availability of element recovery is limited to a few facilities worldwide.
2. The cost of element recovery is higher than that of mining, resulting in economic non-viability.
3. The destruction of both good and bad cells is wasteful when it occurs by undifferentiated destruction.

The cost-effectiveness of cell production processes is limited by the highly compartmentalized value added line. This is happening just because of initially developed production systems and technological design.

3.8 CONCLUSIONS

The conclusion of this chapter explores the fundamentals, design, and manufacturing processes of LiBs. The chemical reactions are also elaborated and explained with the help of basic principles and theories. The battery's primary function is to supply electric current for the operation of the electric drive (motor and gear arrangement) of an EV. Rechargeable lithium-ion batteries (LiBs) are secondary battery types that are environmentally friendly. After this study, researchers will be able to understand the basic fundamentals of LiBs and perform design calculations associated with LiBs.

REFERENCES

1. Chen, Hao, and Julia Shen. "A degradation-based sorting method for lithium-ion battery reuse." *PLoS One* 12, no. 10 (2017): e0185922.
2. Knobloch, Florian, Steef V. Hanssen, Aileen Lam, Hector Pollitt, Pablo Salas, Unnada Chewpreecha, Mark A. J. Huijbregts, and Jean-Francois Mercure. "Net emission reductions from electric cars and heat pumps in 59 world regions over time." *Nature Sustainability* 3, no. 6 (2020): 437–447.
3. Association internationale pour l'évaluation du rendement scolaire. *Global EV outlook 2016: beyond one million electric cars*. IEA (2016).
4. Andwari, Amin Mahmoudzadeh, Apostolos Pesiridis, Srithar Rajoo, Ricardo Martinez-Botas, and Vahid Esfahanian. "A review of battery electric vehicle technology and readiness levels." *Renewable and Sustainable Energy Reviews* 78 (2017): 414–430.
5. Khawaja, Yara, Nathan Shankar, Issa Qiqieh, Jafar Alzubi, Omar Alzubi, M. K. Nallakaruppan, and Sanjeevikumar Padmanaban. "Battery management solutions for li-ion batteries based on artificial intelligence." *Ain Shams Engineering Journal* 14 (2023): 102213.
6. Reddy, Mogalahalli V., Alain Mauger, Christian M. Julien, Andrea Paolella, and Karim Zaghib. "Brief history of early lithium-battery development." *Materials* 13, no. 8 (2020): 1884.

7. Madrid, Christopher, Juan Argueta, and Jordan Smith. *Performance characterization – 1999 Nissan Altra-EV with lithium-ion battery*. Southern California EDISON (1999).
8. Gervillié, Charlotte, Aurélie Boisard, Julien Labbé, Sandrine Berthon-Fabry, and Katia Guérin. "Tin-based materials: the future of anode materials for lithium ion battery?." In *2019 European Space Power Conference (ESPC)* (2019): 5.
9. Mekonnen, Yemeserach, Aditya Sundararajan, and Arif I. Sarwat. "A review of cathode and anode materials for lithium-ion batteries." *Southeast Con 2016* (2016): 1–6. https://doi.org/10.1109/SECON.2016.7506639
10. Thackeray, M. M. *Lithiated oxides for lithium-ion batteries*. National Technical Information Service (1994).
11. Daniel, Claus. "Materials and processing for lithium-ion batteries." *Jom* 60 (2008): 43–48.
12. Xu, Chengjian, Qiang Dai, Linda Gaines, Mingming Hu, Arnold Tukker, and Bernhard Steubing. "Future material demand for automotive lithium-based batteries." *Communications Materials* 1, no. 1 (2020): 99.
13. Stübler, Tillmann, Amine Lahyani, and Amen Allah Zayoud. "Lithium-ion battery modeling using CC–CV and impedance spectroscopy characterizations." *SN Applied Sciences* 2 (2020): 1–8.
14. Kerler, Matthias, Peter Burda, Michael Baumann, and Markus Lienkamp. "A concept of a high-energy, low-voltage EV battery pack." In *2014 IEEE International Electric Vehicle Conference (IEVC)* 2014: 1–8. IEEE.
15. Larminie, James, and John Lowry. *Electric vehicle technology explained*. John Wiley & Sons (2012).
16. Bryden, Thomas S., Borislav Dimitrov, George Hilton, Carlos Ponce de León, Peter Bugryniec, Solomon Brown, Denis Cumming, and Andrew Cruden. "Methodology to determine the heat capacity of lithium-ion cells." *Journal of Power Sources* 395 (2018): 369–378.
17. Crouch, Dell A. "Battery technology for automotive applications." In *Handbook of automotive power electronics and motor drives* (2017): 677–688. CRC Press.
18. Pettinger, Karl-Heinz, Achim Kampker, Claus-Rupert Hohenthanner, Christoph Deutskens, Heiner Heimes, and Ansgar vom Hemdt. "Lithium-ion cell and battery production processes." *Lithium-ion batteries: basics and applications* (2018): 211–226. https://doi.org/10.1007/978-3-662-53071-9_17

Chapter 4

Revolutionizing vehicular braking

The uni-booster integrated disc brake system

Achint Kumar and Gurumukh Das

Department of Mechanical Engineering, Faculty of Engineering, Dayalbagh Educational Institute (Deemed University), Agra, India

4.1 INTRODUCTION

Automotive brake systems are continually evolving, with one of the recent breakthroughs being the introduction of the uni-booster integrated disc brake. This canister-shaped, black component is strategically positioned at the rear of the engine housing, connecting the pedal to the master cylinder. Its underlying principle involves utilizing engine vacuum to amplify the driver's pedal force, ensuring efficient and reliable braking. Modern braking technology, exemplified by the widespread adoption of hydraulic brake booster systems in today's vehicles, has revolutionized how we stop. It empowers heavy and high-speed vehicles to come to a responsive halt, allowing drivers to navigate and avoid potentially dangerous situations with ease. The hydraulic brake booster, exemplified by the uni-booster, has significantly improved safety features and braking capabilities in the automotive industry. The uni-booster's novelty lies in its cutting-edge technology, pushing the boundaries of brake system performance and driver safety.

4.2 LITERATURE REVIEW

In a vehicle, the heat energy generated by the engine is converted into kinetic energy to enable motion. To bring the vehicle to a stop, the process must be reversed, and kinetic energy must be transformed back into heat energy. This transformation is achieved using frictional devices known as brakes. While hydraulic fluid brakes are standard in light vehicles, heavy vehicles often rely on compressed air-powered brakes. Air and hydraulic brakes, as seen in Figure 4.1 (air brake system), are the preferred choices for trucks, where brake safety is paramount. Brake failure in a truck can result in injuries and even fatalities [1].

Advanced technologies, such as the integration of uni-booster and disc brake systems, have significantly improved the performance of air brake systems in the automotive industry. Manufacturers like Volvo Eicher Commercial Vehicles (VECV) and Force Motors have incorporated these

DOI: 10.1201/9781003502470-4

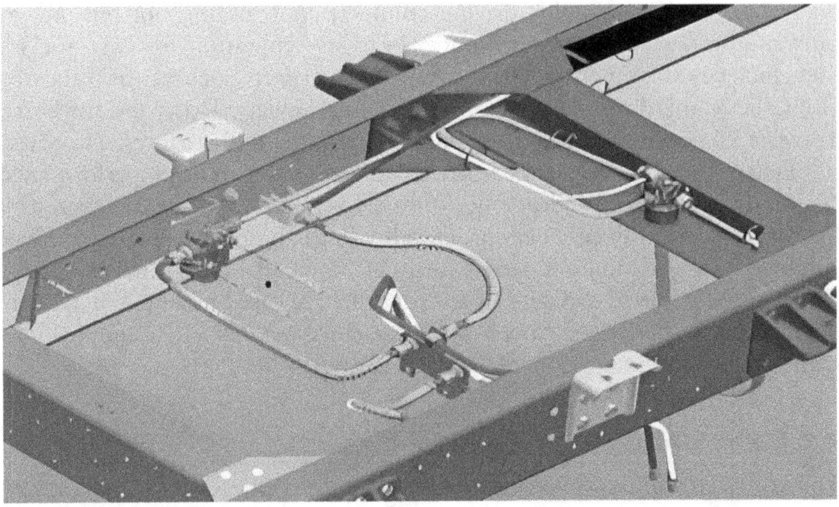

Figure 4.1 Air brake system.

innovations into their vehicles, including the Pro 1095 BSIV model and mono bus. The broader industry trend is adopting a uni-booster integrated disc brake system to enhance vehicular braking. Manufacturers of air brake systems are increasingly exploring the potential of these technologies for future applications [2].

Advancements in hydraulic brakes have also played a crucial role in enhancing braking safety. These developments have now reached a point where they offer cost-effective alternatives to traditional air brakes, making them more accessible and economically viable for truck installations [3].

4.2.1 Objectives

- To provide an efficient and smooth vehicle braking system.
- To enhance safety.
- To identify critical components of the braking system.
- To reduce road accidents involving vehicles.
- To operate effectively in hot environments.
- To explain the operation of a hydraulic power booster (Uni-booster).

4.3 INTRODUCTION TO THE VEHICLE

The Eicher Pro 1095XP stands out as a highly profitable product in the expansive commercial vehicle industry. It falls within a category renowned for its impressive payload capacity and the integration of robust components, ensuring exceptional reliability and longevity. The cabin design emphasizes superior driving comfort, contributing not only to an outstanding driving experience but also to its reputation for best-in-class fuel efficiency. Competing with notable counterparts such as the Tata Ace Gold, the Mahindra Bolero Pickup, the Ashok Leyland Dost, and the Force Traveler 26, each vehicle boasts unique strengths and features. However, the Eicher Pro 1095XP has earned recognition as a mileage performance leader. Its exceptional mileage not only delivers cost savings for operators but also significantly enhances profitability in the highly competitive Indian commercial vehicle market. Key features include best-in-class fuel efficiency, a high-rated payload capacity of 6.9 tons (19 ft), rapid pickup and acceleration, superior cabin comfort, low maintenance with excellent uptime, and high resale value.

4.3.1 Details of the disc brake

The disc brake system, as shown in Figure 4.2 (disc brake rotor, caliper, disc and brake pads), is a mechanism that employs calipers to press pairs of pads against a rotor (or disc), creating friction that retards the rotation of a shaft, such as a vehicle axle. This action either reduces the shaft's speed or holds it stationary and transforms kinetic energy into heat, which must be efficiently dissipated. Hydraulic disc brakes are the most common in motor vehicles. However, the principles of a disc brake apply to nearly any rotating shaft. Disc brakes excel over drum brakes in stopping power due to the disc's superior cooling abilities, making them less prone to brake fade from

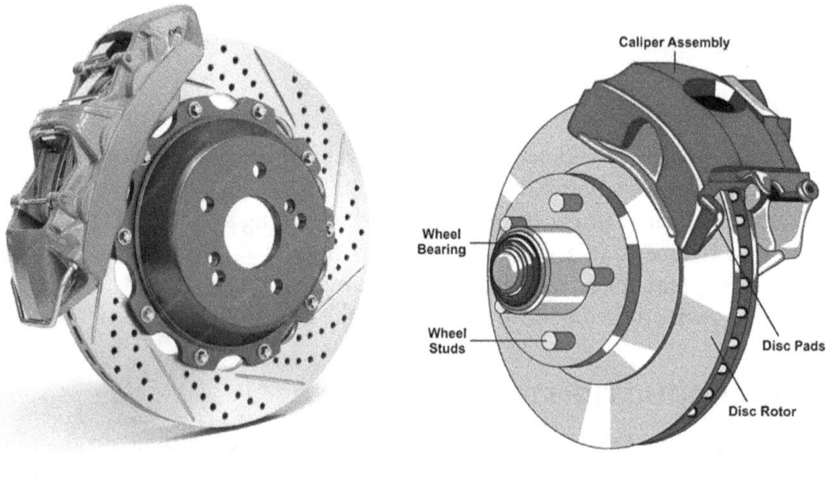

Figure 4.2 Disc brake rotor, caliper, disc, and brake pads.

overheating. They also recover quickly from wet conditions, enhancing their effectiveness.

When the brake is applied, calipers exert pressure on the pads against the rotor, generating friction that gradually slows the vehicle. Over time, the brake pads wear down, and they should be replaced if their thickness falls below a quarter of an inch [4].

The use of disc brakes is on the rise in large and heavy road vehicles, replacing the previously standard drum brakes. Several factors drive this transition. One key factor is that disc brakes do not rely on self-assist, resulting in more predictable brake-force and higher peak brake force. This, in turn, reducing the risk of issues like steering problems or jackknifing, particularly in articulated vehicles. Another advantage of disc brakes is their reduced tendency to fade when exposed to high temperatures. In heavy vehicles, where air resistance, rolling drag, and engine braking play a smaller role in the overall braking force, disc brakes prove more reliable. This is especially crucial compared to lighter vehicles, where drum brakes can experience fading in a single stop [5].

4.3.2 Ceramic brake pads

Ceramic brake pads are dense ceramic material with embedded copper fibers, like fired pottery. Compared to semi-metallic pads, they tend to cause less wear and tear on the rotors. However, one drawback is their reduced effectiveness in colder temperatures or before the brakes have reached their optimal operating temperature.

4.3.3 Need for uni-booster integrated disc brakes

In the past, cars primarily featured drum brakes, eliminating the need for power braking features, as they already possessed several advantages. However, with the widespread adoption of disc brakes, especially on the front wheels, the necessity for power braking systems with integrated boosters became evident. These systems assist drivers in avoiding strenuous exertion when applying the brakes. For individuals who drive frequently or those with limited physical strength, applying excessive pedal force can be challenging and may lead to safety concerns.

Therefore, the brake booster has emerged as a hidden blessing, with the uni-booster circuit shown in Figure 4.3 allowing for easy braking with minimal pedal effort to bring vehicles to a stop when needed [6].

4.4 WORKING PRINCIPLE

The uni-booster integrated disc brake system utilizes two types of hydraulic oil. When the brake pedal is pressed, hydraulic pressure

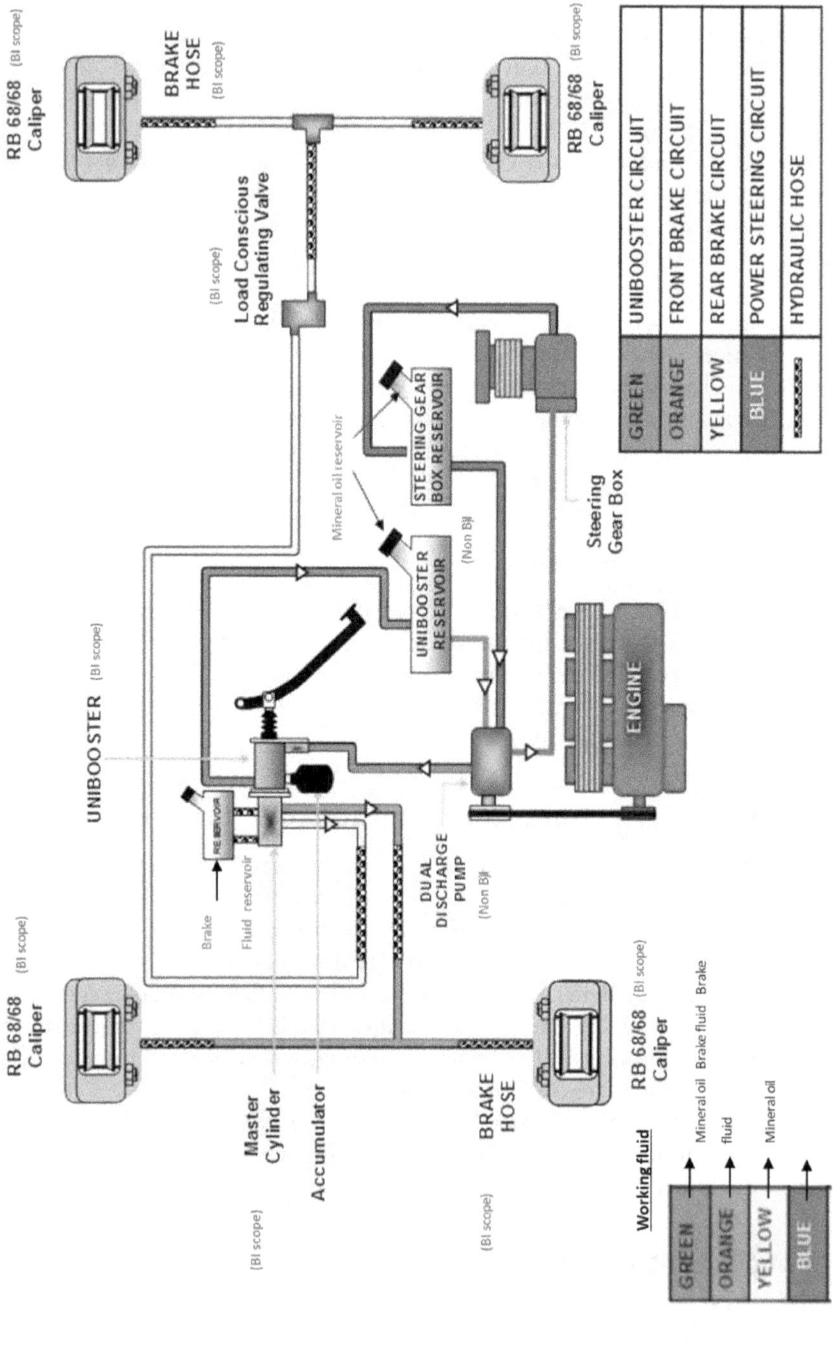

Figure 4.3 Uni-booster circuit.

transmits force to the brakes. However, achieving effective braking requires a force that exceeds what the driver's leg alone can provide. To accomplish this necessary force multiplication, the system employs a combination of mechanical leverage for mechanical advantage and hydraulic force multiplication [7].

4.4.1 Working of disc brake

When a driver applies force to the brake pedal, hydraulic pressure is generated to push the pistons out of their housing. These pistons, in turn, exert pressure on the brake pads, creating friction against the rotating disc faces and effectively slowing down the vehicle.

Hydraulic fluid plays a crucial role in this process by transmitting the pressure to the floating pistons on both sides. When the driver releases their foot from the brake pedal, the hydraulic pressure on the friction pads is released, allowing the pistons to retract and disengage from the disc [8].

4.5 NEED FOR NEW PRODUCTS

The introduction of a new product, as shown in Table 4.1, is imperative due to the evolving dynamics of the commercial vehicle market. Major challenges, including BSVI compliance, 5T optimization, and the increasing demand for customized application vehicles, necessitate strategic decisions. Despite these obstacles, the market presents unique opportunities, such as a growing trend among younger customers seeking comfort through next-gen cabins and a rising interest in alternative fuels.

Addressing these challenges requires strategic decisions, including market forecasts for volume evaluation, benchmarking vehicle finalization, and OEM customization of ready-built truck alternatives. Moreover, meeting consumer preferences involves aligning product requirements with key components like cabin, engine, and freight body. To develop a successful next-gen product, it is essential to adapt to evolving transportation patterns like the Hub & Spoke model and navigate regulatory changes such as GST adoption and emission criteria shifts (BSIV to BSVI).

In a dynamic market with issues demanding innovative solutions, the introduction of a new product is crucial. A successful launch in the 4-9T GVW segment, both domestically and internationally, requires an understanding of client preferences, responsiveness to regulatory changes, and capitalization on emerging trends. The new product requirement, as given in Table 4.1, plays a pivotal role in guiding development to meet market demands and stay ahead of industry challenges.

Table 4.1 New product requirement

What is happening	*Why it is happening*	*Gap analysis*	*Action plan*
• Lower average vehicle life reducing resale value (8~10yrs) • Adoption of alternative fuels (CNG, Electric, Biodiesel) • Next-gen product introduction from OEMs (TML-U, ALL, Boss, Guru ALL-NISSAN Partner)	• Ban on 10 years old diesel vehicle due to environmental concerns • Change in emission norms BSIV (2017) and BSVI (2020) • Adoption of platform proliferation strategy dual product options	**Strength** • Leadership-FE, TAT Payload Best fit for rated load • application-Parcel & Courier, E-commerce, F&V etc. • Spacious and comfortable cab • Optimum power and torque	• New platform strategy for Pro 1000 Successor (4-9T GVW) • Market study to understand segment dynamic of 4-9T GVW domestic & export • New market exploration capturing trends, platform strategies regulations sufficing Euro emissions
• High acceptance of next-gen cabin due to changing customer base and rising education level • Customer preference for customized and ready-to-use • Emerging Hub & Spoke Transportation Model • Better reach with last-mile distribution	• Younger customer base seeking comfort along with hygiene factor • Ready-built truck options by OEMs • GST Implementation • Smart city development. Swachh Bharat & enhance road infra	**Challenges** • Readiness for BSVI • Optimized product for the 5T category • Customized application vehicle • Narrow cabin for inner city/narrow lane operation Challenge in Overload application • Suitable engine for addressing smaller tonnage (GVW<5T)	• Market forecasting volume assessment and benchmark vehicle finalization • Defining product requirement as per primary aggregates – Cabin, Engine, and cargo body • Pre-requisite and BPS refinement and agreement with key stakeholders

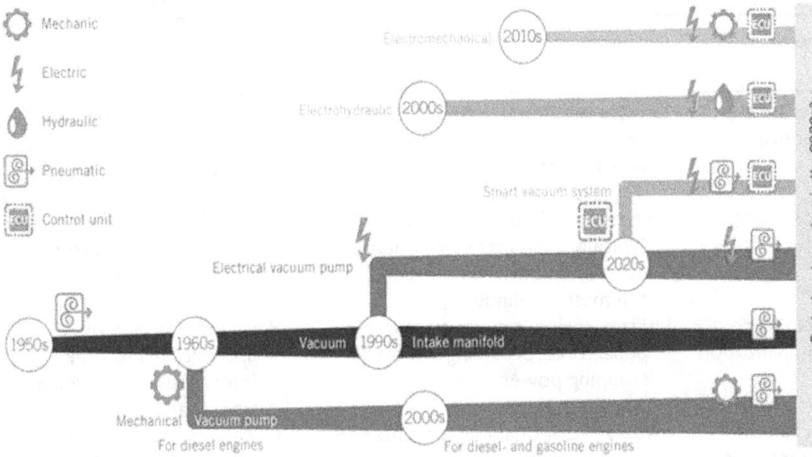

Figure 4.4 Guideline for enhancing brake booster systems.

4.5.1 Strategy alignment

The current state of the 5- to 9-ton commercial vehicle industry comprises approximately 50,000 units as of 2021, with a consistent average growth rate of around 10% during that year. Anticipating evolving factors such as urbanization, innovative urban development initiatives, cleanliness campaigns, and urban access regulations, the light-duty truck (LDT) industry's volume is expected to increase significantly, reaching approximately 55,000 units by 2024.

To respond to market demands, the product range has been comprehensively renewed, including the introduction of the Pro 1000 series, with a capacity of 4 to 14 tons, and the Pro 3000 series, designed for 13 to 16 tons. Plans are underway to develop a successor for the Pro 3000 series, catering to the 10 to 16 ton gross vehicle weight (GVW) segment. A strategic platform is also being defined to strengthen the brand's position in India and international markets for 4 to 9 tons LDTs, with implementation expected in 2020 and beyond. Additionally, the *Guideline for Enhancing Brake Booster Systems*, as depicted in Figure 4.4, is being incorporated to ensure optimal braking performance and safety in new product developments.

4.5.2 Uni-booster brake system versus conventional friction brake system

The technical comparison tabulated in Table 4.2 highlights the key differences between the uni-booster brake system and the conventional

Table 4.2 Uni-booster brake system versus conventional friction brake system

Aspect	*Uni-booster brake system*	*Conventional friction brake system*
Operating principle	Utilizes engine vacuum to amplify pedal force, improving braking effectiveness.	Operates on friction between brake pads and a rotor or drum to slow down the vehicle.
Booster type	Employs a uni-booster technology, typically a circular component positioned between the pedal and the master cylinder.	Lacks a booster; pedal force is directly applied to the master cylinder.
Brake force amplification	Utilizes engine vacuum to multiply pedal force, providing enhanced stopping power.	Pedal force is directly proportional to brake force, with no additional amplification.
Mechanical complexity	Involves additional components like the uni-booster, check valves, and vacuum lines.	Generally simpler in design, with minimal additional components.
Brake response time	Offers faster and more responsive braking due to the amplification of pedal force.	Response time may be slightly slower in comparison to systems with boosters.
Brake fade resistance	Exhibits improved brake fade resistance due to the boosted braking power.	May experience brake fade under heavy or prolonged braking.
Brake maintenance	May require specialized maintenance for the uni-booster and associated components	Standard brake maintenance and servicing with no booster-specific requirements.
Weight and space requirements	Typically, it adds some weight and occupies additional space within the brake system.	Compact and lightweight, with minimal impact on vehicle weight and space.
Application	Primarily used in modern light commercial vehicles (LCVs) and trucks.	Widespread use across various types of vehicles, including passenger cars and LCVs.
Market adoption	Gaining popularity for improved braking in the automotive industry.	A well-established and widely used brake system in the automotive market.

friction brake system. This analysis delves into operational principles, mechanical intricacies, and performance features of these two braking systems, enhancing our comprehension of their respective applications and adoption within the automotive industry.

4.6 COMPETITOR ANALYSIS

The Indian light-to-medium duty (LMD) truck market is highly competitive, with numerous domestic and international players vying for a significant

Table 4.3 Competition landscape

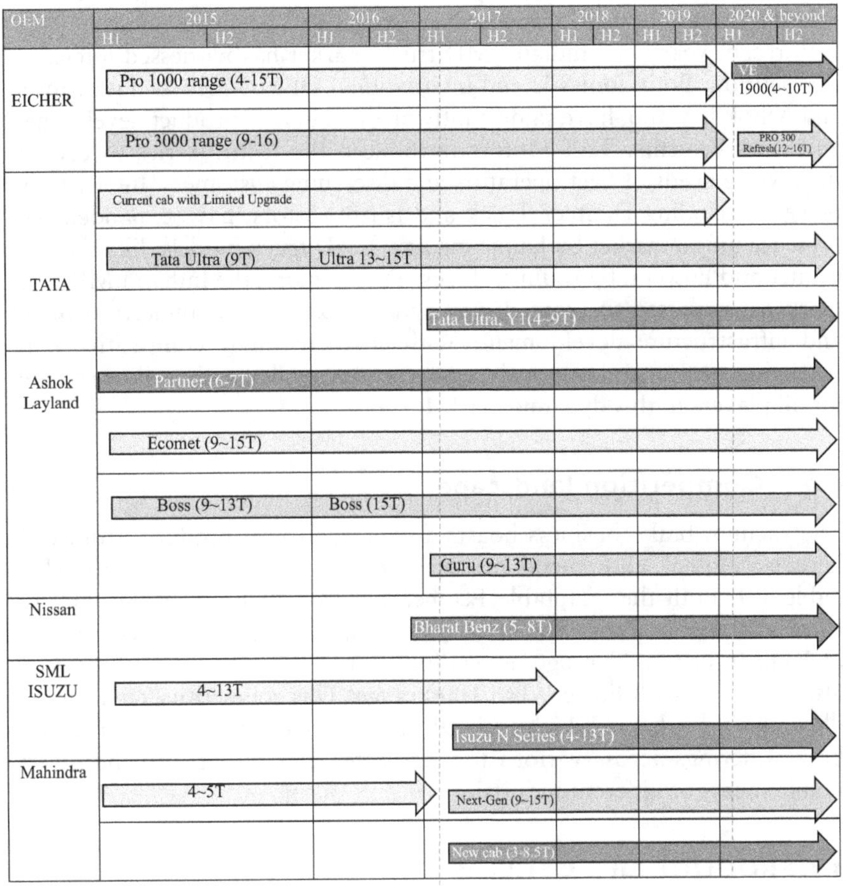

market share. Notable domestic original equipment manufacturers (OEMs) include Tata Motors, which dominates the market with a commanding share of over 40%. Tata Motors offers a wide range of LMD trucks under the 'Ultra' brand, including models like the Ultra T.7, Ultra T.9, and Ultra T.11. Following closely is Ashok Leyland, the second-largest player in the Indian LMD truck segment, maintaining a market share exceeding 20% and providing LMD trucks under the 'Dost' and 'GURU' brands. On the international front, global heavyweight Daimler Truck has established a significant presence in the Indian LMD truck market, introducing a variety of LMD trucks under the 'Fuso' brand, such as the Fuso Canter and Fuso FE. Isuzu Motors, another prominent global player, is also a key participant in the Indian LMD truck market, with its 'N series' brand featuring models like the NQR and NPR. Beyond these industry leaders, several other international OEMs, including Hyundai, Hino, JAC, JMC, and Foton, are closely monitoring the Indian LMD truck market, poised to introduce their product ranges soon [9].

In recent years, the Indian LMD truck market has witnessed intensified competition. Both domestic and international OEMs, such as Tata Motors and Ashok Leyland, have significantly invested in new product development and enhancements. Tata Motors introduced the 'Ultra' series to cater to the diverse needs of fleet operators and discerning customers. International players, including Daimler Truck and Isuzu Motors, have expanded their presence in the market by launching new LMD truck models, like the Fuso Canter and Fuso FE by Daimler Truck. Furthermore, the Indian LMD truck market is expected to grow, driven by urbanization, e-commerce demand, and infrastructure development. With an increasingly competitive landscape, innovation and responsiveness to customer demands will be essential for all players in this dynamic market.

4.6.1 Competition landscape

The vacuum brake business boasts a wide variety of products from well-known original equipment manufacturers (OEMs), as represented in Table 4.3. Both the adaptable Ecomet line from Ashok Leyland and the Ultra series from Tata, which focuses on the 9T market, are competitors of Eicher's Pro 1000 range, extending from 4 to 15T. Tata's Ultra series also competes with Eicher's Pro 1000 range. This competitive environment illustrates a market in which each OEM strategically distributes its vacuum brake offerings across various tonnage categories to cater to the diverse requirements of different industries.

4.7 PRODUCT LIFE CYCLE

Manufacturers typically employ a product refresh strategy every five years after starting production [10] as shown in Figure 4.5. In this strategy, the base variant is phased out, and the mid variant becomes the new base variant. Simultaneously, the high variant assumes the role of the mid variant in subsequent refresh programs. Each refresh cycle introduces a new high variant. The base variant is designed to meet essential requirements with standard features. In contrast, the high variant is tailored to satisfy more advanced customer needs, and the mid variant offers a well-balanced combination of fundamental and delightful features.

4.7.1 Market selection

The manufacturer's LMD products play a crucial role in achieving significant export volumes, contributing to overall sales [11]. To identify export

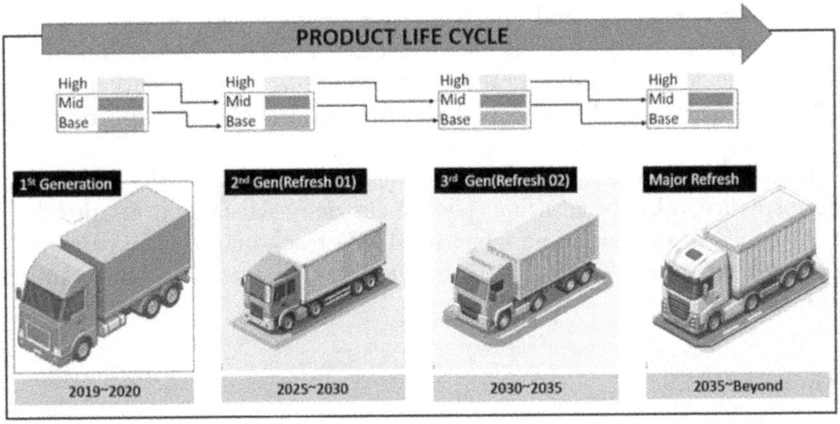

Figure 4.5 Product life cycle.

opportunities, segment practices, applications, product specifications, features, and other relevant factors for export prerequisites, a comprehensive market study is necessary.

Qualification criteria for LDTs for international business purposes include the following:

- The CBC curb weight must be under 2.5 tons (Dubai).
- The truck's GVW should be below 6 tons (Oman and Kuwait).
- The DSD payload must be at least 3 tons (Qatar).
- If the load body height is reduced, it will meet Bahrain's requirement, as the total unladen weight will be less than 3 tons.

4.8 CONCLUSIONS

This study emphasizes the significance of designing and developing a uni-booster integrated disc brake system for the manufacturing of LCVs. Implementing this innovation is poised to offer substantial customer benefits, delivering superior quality and performance within the brake systems of LCVs. This advancement is well-positioned to address the diverse needs of customers, ultimately leading to elevated customer satisfaction.

In the fiercely competitive Indian LMD truck market, major players like Tata Motors, Ashok Leyland, Daimler Truck, and Isuzu Motors are making substantial investments in product development to cater to customer requirements and maintain market share. International OEMs, such as Hyundai, Hino, JAC, JMC, and Foton, are closely monitoring this dynamic market, preparing to introduce their product ranges.

As the LMD truck market continues to expand due to urbanization, e-commerce demand, and infrastructure development, innovation and responsiveness to customer demands are crucial. The introduction of the uni-booster integrated disc brake system aligns with these market trends, offering superior quality and performance to meet the evolving needs of customers.

In conclusion, this innovation is positioned to enhance the safety and performance of LCVs, ensuring elevated customer satisfaction. Its introduction comes at a crucial time when the LMD truck market is experiencing heightened competition and rapid growth, making it a vital addition to the industry.

REFERENCES

[1] H. Seiff, "Heavy truck brake adjustment-problems and solutions," in *Road Transport Technology, Proceedings of the International Symposium on Heavy Vehicle Weights and Dimensions*, 1992, pp. 575–582.

[2] A. Kumar, V. Kumar, V. Modgil, and A. Kumar, "Stochastic Petri Nets modelling for performance assessment of a manufacturing unit," *Mater. Today Proc.*, vol. 56, pp. 215–219, 2022, doi: 10.1016/j.matpr.2022.01.073.

[3] R. L. Hoover, S. B. Zagorski, and T. Van Buskirk, "Air disc and S-cam brake stopping performance comparison for class 8 tractor and trailer combinations," in *SAE Technical Papers*, Nov. 2005, no. 724, doi: 10.4271/2005-01-3614.

[4] M. Knowles, "Through-life management of electric vehicles," *Proc. CIRP*, vol. 11, pp. 260–265, 2013, doi: 10.1016/j.procir.2013.07.074.

[5] L. Wang and H. Zheng, "A slip-rate-based braking force distribution algorithm for the electronic braking system of combination vehicle," in *SAE Technical Papers*, Sep. 2014, vol. 2014-Janua, doi: 10.4271/2014-01-2385.

[6] H. Leffler, E. Petersen, and B. Shilton, "A new Anti-Lock-System (ABS) for light commercial vehicles and passenger cars with integrated hydraulic brake booster," in *SAE Technical Papers*, 1987, No. 865140.

[7] L. Zhang, C. Ruan, and D. Meng, "The influence of vacuum booster design parameters on brake pedal feel," *SAE Int. J. Passeng. Cars – Mech. Syst.*, vol. 7, no. 4, pp. 2014-01–2499, Sep. 2014, doi: 10.4271/2014-01-2499.

[8] S. Yang *et al.*, "Studies on brake pedal feeling based on a novel mechatronic booster," Apr. 2016, doi: 10.4271/2016-01-0014.

[9] H. Liimatainen, O. van Vliet, and D. Aplyn, "The potential of electric trucks – an international commodity-level analysis," *Appl. Energy*, vol. 236, pp. 804–814, Feb. 2019, doi: 10.1016/j.apenergy.2018.12.017.

[10] A. Kumar, H. Singh, P. Kumar, and B. AlMangour, *Handbook of Smart Manufacturing*. Boca Raton: CRC Press, 2023.

[11] S. Rani, K. Tripathi, and A. Kumar, "Machine learning aided malware detection for secure and smart manufacturing: a comprehensive analysis of the state of the art," *Int. J. Interact. Des. Manuf.*, Oct. 2023, doi: 10.1007/s12008-023-01578-0.

Chapter 5

Analyzing the impact of bumper height on pedestrian injuries using explicit dynamics

Abhishek Agarwal,[1] *Rafael Cavicchioli Batista,*[2] *and Adash Gurung*[1]

[1]Department of Mechanical Engineering, College of Science & Technology, Royal University of Bhutan, Chukka, Bhutan
[2]Department of Mechanical & Manufacturing Engineering Technology, Vermont State University, Vermont, USA

5.1 INTRODUCTION

The transportation problem is a classic optimization problem that has numerous applications in manufacturing industries. The radiator, front grille, or bonnet of a car may be damaged by low-speed collisions. Through reducing the collision energy, bumpers can shield these important parts. The effectiveness of the bumper for shielding those parts may be considerably enhanced with the use of proper design and elements. Bumpers may boost road grip and downward drag in addition to providing security, which might enhance vehicle handling at high speeds. Figure 5.1 shows the front bumper of sedan car.

Bumpers are not designed to improve vehicle crashworthiness or protect occupants in collisions. However, they can help to protect headlamps, rear lamps, and other parts from damage during low-speed impacts. Bumpers are made of a heavy steel beam that is attached to impact absorbers. The impact absorbers are designed to deform and absorb the energy of the impact, which helps to reduce damage to the bumper and other parts.

5.1.1 Anatomy of a pedestrian-vehicle crash

Figure 5.2 shows the crash zones for a four-wheel vehicle. As it can be seen, the most common collision points are the bonnet, the driver's side door, and the passenger's side door. The rear of the vehicle is the least likely to cause injuries or fatalities. Based on the US National Highway Traffic Safety Administration data, schematic image in Figure 5.2 shows that, out of all the collision zones that are frequently encountered in a four-wheel vehicle, the front zones account for the majority of pedestrian injuries, with an estimated 71.1% of all injuries [1].

DOI: 10.1201/9781003502470-5

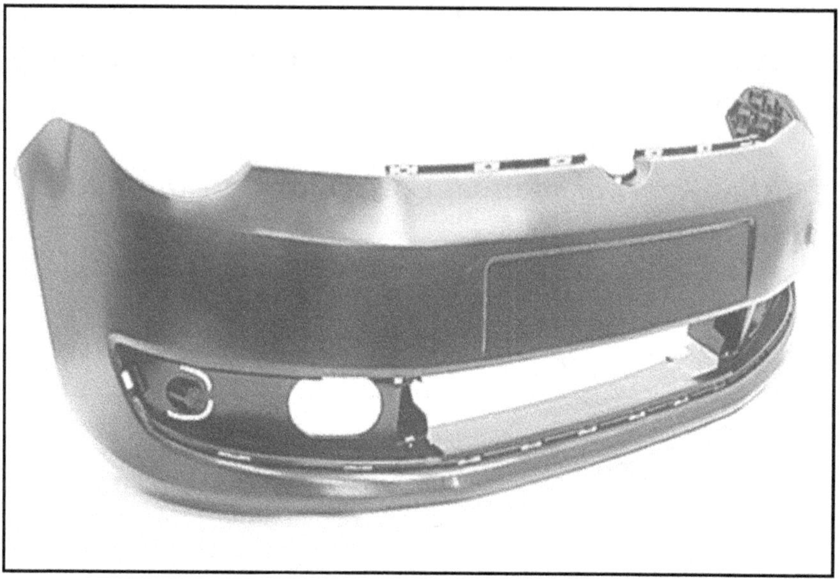

Figure 5.1 Front bumper of a sedan car.

Figure 5.2 Front impact zone of four-wheel vehicles.

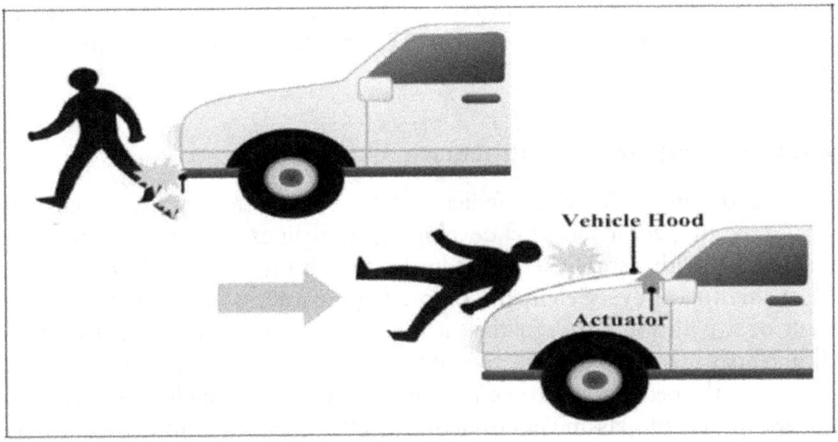

Figure 5.3 Sequence of events in a pedestrian–vehicle crash [2].

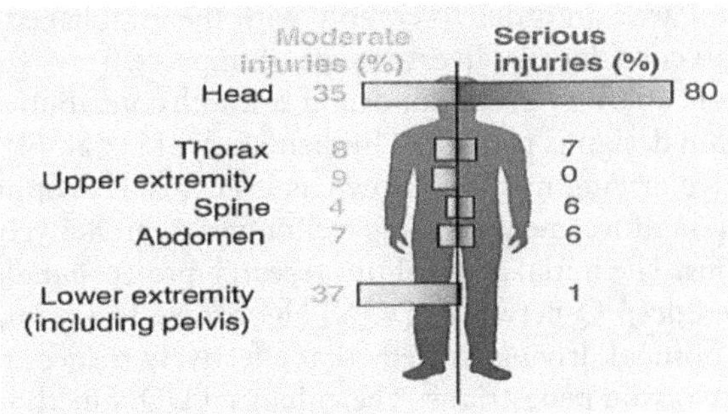

Figure 5.4 Injury distribution in crashes involving pedestrians [3].

The front structures of a vehicle are responsible for most of the critical injuries to pedestrians. The lower limbs of the pedestrian typically make contact with the bumper first, followed by the pelvis and then the head. Figure 5.3 shows the crash sequence that occurs during a collision.

The pedestrian's head collides with the bonnet and may reach the windshield. After the collision, the pedestrian may slide on the road or be thrown onto the bonnet as the vehicle continues to move forward before coming to a complete stop.

5.1.2 Injury profiles and countermeasures

The upper body region above the pelvis and the head are the most affected body parts in a collision. Head injuries are the most common cause of fatalities in pedestrian accidents. Figure 5.4 shows the injury distribution chart for pedestrian accidents.

The chart shows that the head region is the most affected, with 80% of injuries being serious. The thorax region has the fewest serious injuries, at 7%. The region below the pelvis has 37% moderate injuries, but only 1% fatal injuries.

5.1.3 Evaluations of countermeasures

The European Enhanced Vehicle-Safety Committee (EECV) ensures the safety of pedestrians by developing guidelines for crash testing. In accordance with established standards, the testing protocols necessitate the evaluation of every element comprising the automotive structure in the event of a collision between the car and a pedestrian at a velocity of 40 kilometers per hour. The objective of this evaluation as shown in Figure 5.5 is to assess the pedestrian protection capabilities of the vehicle across various age groups, encompassing children and adults. The evaluation will consider the specific demographic characteristics associated with each age group. It is necessary to adjust the testing conditions to accommodate the diverse impact characteristics that arise from alterations in the overall design of the vehicle's frontal area, in order to comply with the established standards. It is imperative to consider the diverse range of impacts.

In this section, an overview of important research contributions in the field of bumper beam design is provided. In their study, Li et al. [4] introduced a new cooperative driving method known as L-HMC. This approach utilizes advanced deep reinforcement learning techniques to proactively mitigate the risk of collisions. The initial stage of this research project entails the creation of an enhanced deep Q-network (DQN) algorithm. The primary goal is to determine an optimal driving strategy that effectively reduces the likelihood of accidents involving pedestrians. The enhanced DQN method utilizes two separate replay buffers that store conflicting data in order to accelerate the learning process of identifying the most effective driving strategy. Following

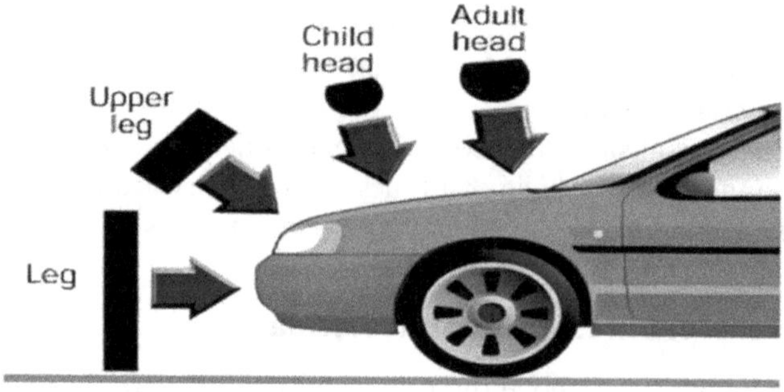

Figure 5.5 Impactor test by EECV to assess the pedestrian protection capabilities [3].

this, a suggested approach involves the implementation of a human–machine collaborative driving system. This system is designed to support human drivers in their adherence to established driving policies. The main goal of this system is to reduce the likelihood of pedestrian accidents caused by unsafe driving behaviors exhibited by human drivers. The effectiveness of the human–machine cooperative driving strategy is assessed on the PreScan simulation platform using a dynamic model of an actual vehicle. This measure facilitates the optimization of the system's precision. The outcomes show that a driving policy for avoiding pedestrian collisions may be learned using a deep reinforcement learning approach with a quick convergence rate. The primary materials of interest in the research by Sudirja et al. [5] are Mild steel S45C and Vinyl ester with Microsphere (CRVeM) material. When used as a constituent material for the fascia bumper, CRVeM exhibits an enhanced capacity for energy absorption prior to experiencing irreversible degradation. This phenomenon persists despite the fact that the material does not contribute much to the overall mass of the bumper component. In contrast to the aforementioned, the utilization of mild steel S45C is seen, serving as the principal constituent of the front bumpers in LIPI electric automobiles, and is effectively applied inside these automotive vehicles. In low-speed crashes, vehicles have a notable ability to undergo deformation, hence enhancing the safety of pedestrians. The mechanical properties and behavior of each material were investigated via the use of thin-walled tubes in flexural testing and crash simulations. Based on the outcomes obtained from the ABAQUS UNIFIED finite element analysis (FEA) simulation, the researchers have deduced that the CRVeM material exhibits a higher level of deformation in comparison to the mild steel S45C material. Specifically, the CRVeM material shows a strain of 24.4% relative to its original length. This finding has positive implications for pedestrian safety and the mitigation of low-speed events, particularly in parking lot settings involving motor vehicles.

Baek et al. [6] offered a method for creating a multisensor-based, wireless vehicular communications-based vehicle collision warning system. The study performed a thorough integration of radar, lidar, camera, and wireless vehicle communication data to deliver the driver with timely collision warnings and facilitate the prediction of trajectories of distant objects. Furthermore, this measure was implemented with the objective of providing the driver with the ability to anticipate potential collisions. The researchers devised and assessed the proposed vehicle collision system within simulated driving environments, encompassing scenarios featuring vehicle-to-vehicle collisions and vehicle-to-pedestrian collisions. Furthermore, these scenarios encompassed occurrences entailing collisions between motorized vehicles. Zhang et al. [7] offered a method for creating a multisensor-based, wireless vehicular communications-based vehicle collision warning system. This study involved the extensive integration of radar, lidar, camera, and wireless vehicle communication data. The main goal was to deliver the driver with prompt collision warnings and the capability to predict the paths of distant objects. Furthermore, this action was executed with the objective of providing the driver with the ability to anticipate potential collisions. The researchers devised and assessed the proposed vehicle collision system within simulated driving environments, encompassing scenarios encompassing both vehicle-to-vehicle and vehicle-to-pedestrian collisions. Moreover, these settings encompassed occurrences of vehicular collisions. The quantitative findings indicate that the newly suggested wearable airbag demonstrates promise in providing effective protection for pedestrians by substantially decreasing the occurrence of head and chest injuries. Due to the significant impact of brain injuries, their severity is evaluated through a comprehensive range of indicators of damage. The metrics encompassed intracranial pressure, acceleration at 3 milliseconds, and HIC15. Traumatic brain injuries are regarded as the most severe form of injuries. A research investigation is undertaken to analyze the influence of different variables, such as the mass of the inflator gas and the length of the pulling strap, on the protective abilities of the wearable airbag. The objective is to improve the safety characteristics of the airbag. Based on the results of the parametric analysis, it has been determined that the most effective method for safeguarding the head of the HBM entails employing a 20 mm pulling strap in conjunction with 26 g of gas.

According to Senthilnathan et al. [8], the fuzzy method of analytical hierarchy (AHP) or the fuzzy method for ranking desires by the similarity of ideal solutions (TOPSIS) were combined to create a hybrid multicriteria decision-making analysis (MCDA) method. It was found that carbon fiber reinforced in an epoxy matrix is a better material for automotive bumper beams. This investigation will help designers and product development staff to choose the best materials for the automotive industry. Schulz et al. [9] suggested utilizing a latent-dynamic conditional random field approach, provide a novel method for sophisticated video-based driving assistance systems to recognize pedestrian intentions. The methodology utilizes data acquired from a stereo-video system to detect pedestrians and evaluate the localization of human heads. This facilitates the incorporation of pedestrian dynamics into the methodology while upholding situational awareness. The model has the capacity to incorporate both extrinsic and intrinsic manifestations of class dynamics. In order to evaluate the performance of our system, we utilize a publicly accessible dataset that consists of instances where individuals engage in lateral maneuvers, such as curbside stops, road turns, and road crossings. The experimental findings presented by the authors offer proof that the proposed methodology outperforms current state-of-the-art techniques in accurately determining pedestrians' intentions. It demonstrates enhanced performance in terms of both stability and differentiation between classes.

Luo et al. [10] researched preventing glare interference, a flexible high-beam regulating mechanism is used. The light source utilized in the adaptive high beams is comprised of an assemblage of light-emitting diode (LED) particles. The luminous intensity distribution of a single LED can be determined by analyzing the lighting zones of a cluster of LED particles. The precise calibration of a vehicle's headlights or taillights with the images recorded by its on-board camera allows for the identification of other moving vehicles that are traveling in the same direction. The LED light zones on the observed vehicles have been appropriately designated. The adaptive high beam management system employs a selective deactivation mechanism for LED particles in specific vehicle zones, while ensuring that the remaining LED particles continue to provide illumination. The proposed approach entails the incorporation of a lamp controller, an image processor, and controlled LED particles to establish an adaptive high-beam control system. This technique exhibits a significant level of novelty and originality. The implementation of the suggested adaptive high lights would result in a reduction in the intensity of glare disturbance, thereby improving the safety of driving during nighttime.

Singh et al. [11] proposed an innovative framework for the automated detection and identification of traffic accidents. This framework utilizes surveillance video as the primary source of data. Instead of using traditional handcrafted features, the proposed method involves dynamically learning feature representations from the spatiotemporal volumes of raw pixel intensity. The incidence of a vehicular collision is noteworthy given its atypical characteristics. The existing methodology showcases the ability to extract extensive information through the utilization of denoising autoencoders that have been trained on regular traffic footage. The evaluation of accident probability involves the consideration of both the reconstruction error and the deep representation probability. During the training phase of an unsupervised approach, a single-class support vector machine (SVM) is employed to calculate the probability of the deep representation. Moreover, the utilization of the vehicle's path intersections is implemented to improve the system's reliability and minimize the occurrence of false alarms. In research conducted by Lanzaro et al. [12], an evaluation was conducted to analyze the potential risks linked to interactions between vehicles and pedestrians in four major urban regions across the United States and Asia, specifically Boston, Las Vegas, Pittsburgh, and Singapore. The researchers employed the peak over threshold methodology, which is based on extreme value theory, to analyze the data and assess the corresponding risks. The integration of multiple variables into a Bayesian hierarchical model is widely recognized, allowing for the evaluation of risk related to individual interactions. It makes use of a sizable AV database. Due to their numerous sensors, AVs are able to continuously record data regarding their surroundings, particularly the locations as well as speeds of other drivers. Outcomes indicate that there are considerable differences in risk between cities.

Vaduka et al. [13] observed that the bumper serves a crucial function in reducing the impact forces generated during a car collision. The presence of it is essential for effectively addressing these requirements. The design of the front or rear section of a vehicle is specifically engineered to endure impacts while maintaining the structural integrity of the interior safety systems. In current discussions, the fuel efficiency of an automobile holds considerable significance. The automotive bumper is regarded as a critical component due to its substantial size and weight. The main goal of impact research is to enhance the composition of bumper materials. This improvement would result in a reduction of impact forces generated during crashes, leading to a mitigation of the resulting consequences. Furthermore, this will allow us to select the material with the utmost capacity for assimilation. The analytical methodology consists of various components, such as the conceptualization of the bumper design, the utilization of Autodesk Inventor software, and the integration of the bumper into Ansys 19.2 for numerical assessments. In order to evaluate the structural soundness of the bumper, it is imperative to obtain three distinct velocity measurements, specifically 80, 100, and 120 kilometers per hour. During the implementation of the experimental methodology, composite materials such as S-glass epoxy and carbon fiber reinforced polyetherimide are employed to augment the material's mechanical strength. Further occurrences of composite materials can be observed. After conducting a thorough analysis of the potential impacts, the decision was made to produce the bumper utilizing the high-quality fabric. Basith et al. [14] state that automobile fenders are integral components that serve a critical function in mitigating the impact force experienced during collisions and minimizing harm to both the vehicle and its occupants. The function in question plays a crucial role in ensuring the overall safety of the vehicle. Significant advancements have been made in the composition and design of bumpers for passenger automobiles in recent years. During the design stage of the bumper, a considerable focus was given to the evaluation of the correlation between material cost and weight ratio. The main objective was to achieve cost-effectiveness and minimize maintenance expenses for the bumper. The aim of this research is to evaluate the efficacy of chromium-coated mild steel, aluminum alloy, and thermoplastic materials in three distinct testing scenarios, each performed at different velocities. The objective of this investigation pertains to the fender of a passenger vehicle. The vehicle's velocity undergoes a gradual increase until it reaches its maximum speed. The initial velocity measurement was documented as 40 kilometers per hour. The initial value was subsequently increased to 60 kilometers per hour, and then further raised to 80 kilometers per hour. Following a thorough investigation into body deformation, strain, and tension at various velocities, the optimal material for the automobile's bumper was selected based on the established criteria. Analytical software is utilized to detect and evaluate flaws or deteriorations, while computer-aided design (CAD) software is employed to construct three-dimensional models. Kumar et al. [15] stated the commonly used abbreviation for the fourth industrial revolution is '4.0 industry'. Industry 4.0 pertains to the process of digitalizing and transforming industrial and value-creation procedures. The concept of 'Industry 4.0' pertains to the ongoing integration of sophisticated and intelligent technologies for the purpose of automating conventional industrial and manufacturing processes. The main objective of integrating machine-to-machine (M2M) operations within the Internet of Things (IoT) framework is to improve automation, communication, and self-monitoring capabilities. This integration also enables the development of intelligent machines with autonomous problem-solving and analytical abilities [16]. Goyat et al. [17] have presented a book based on intelligent materials. The book discussed various manufacturing processes involved in the development of intelligent materials. The capabilities, limitations, and distinct characteristics of intelligent materials are presented. Haleem et al. [18] have presented the latest advancement in the field of additive manufacturing. The work presents different stages of manufacturing, prototyping right from the material selection to final product completion, which serves as a reference for establishing manufacturing principles and guidelines in additive manufacturing. Table 5.1 summarizes the important literature reviews.

The primary purpose of the bumper is to minimize the effect of impact on engine and other vital components at low speeds. The severity of pedestrian injuries is significantly influenced by bumper height and therefore needs investigation. The crash analysis of a bumper with a pedestrian can be performed using a legform impactor model and in accordance with the

Table 5.1 These research works have contributed to enhancing the understanding in analyzing the impact of bumper height on pedestrian injuries using explicit dynamics

Author	*Objective function*	*Front impact*	*Side impact*	*Energy absorption*
Li et al. [6]	Crash Analysis	✓		
Sudirja et al. [7]	Crash Analysis			✓
Baek et al. [8]	Crash Analysis	✓		
Zhang et al.[9]	Collision System			✓
Senthilnathan et al. [10]	Crash Analysis		✓	
Schulz et al. [11]	Multi objective	✓		
Luo et al [12]	Multi objective		✓	
Dinesh Singh et al. [13]	Multi objective	✓		
Gabriel et al.[14]	Multi objective		✓	
Vaduka et al.[15]	Multi objective			✓
Basith et al.[16]	Multi objective		✓	
Kumar et al. [17]	Multi objective			✓
This study	Multi objective	✓	✓	✓

European Enhanced Vehicle-safety Committee (EEVC) regulations. The new material known as Al 921T can also be tested. The current research looked into how low-impact crashes with bumpers harm pedestrians. Investigations were also done into how bumper height and material affected harm. The actions were as follows:

- The legform and front bumper of the car were created as CAD models. The legform was created using an EEVC-compliant design.
- LS-DYNA was used to do an explicit dynamic evaluation of the collision for bumper elevations of 210, 260, and 310 mm.
- The crash study resulted in the determination of the energy absorption curve, deformation, or equivalent stress.
- Al921T was used as the new bumper material, as well as the crash simulation was redone.
- The results of the investigation could be applied to lessen pedestrian injuries and enhance the layout of automobile bumpers.

Section 5.1 reviews the background information and existing literature on the topic. It discusses the work of other researchers in this area, and identifies the gaps in knowledge that the present study aims to fill. It also states the problem that the study is addressing and the objectives of the study and defines the key terms that will be used in the study. Section 5.2 describes the methodology that was used in the study. It discusses the data that was collected, the methods that were used to analyze the data, and the limitations of the study. Section 5.3 presents the results of the study. It discusses the findings of the study, and how they contribute to the existing knowledge on the topic. Section 5.4 summarizes the main findings of the study and discusses the future scope of the research. It identifies the questions that remain unanswered and the directions for future research.

5.2 MATERIAL, DESIGN, AND MODELING APPROACH

In order to examine the effects of bumper height and material characteristics on pedestrian injuries in low-impact crashes, this study takes a methodical approach. The investigation combines explicit dynamic evaluation with the use of CAD modeling to thoroughly evaluate the energy absorption, deformation properties, and equivalent stress distributions related to different bumper heights [19]. Additionally, the possibility of Al 921T, when used as a bumper material, could improve pedestrian safety is thoroughly investigated. The research approach is displayed in the flowchart provided by Figure 5.6.

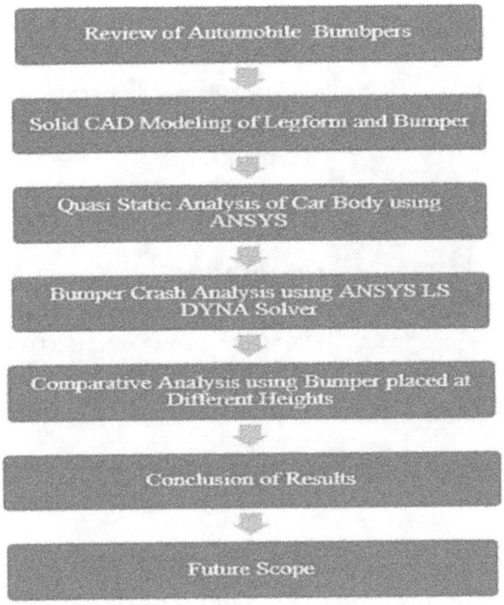

Figure 5.6 Flowchart of the proposed study.

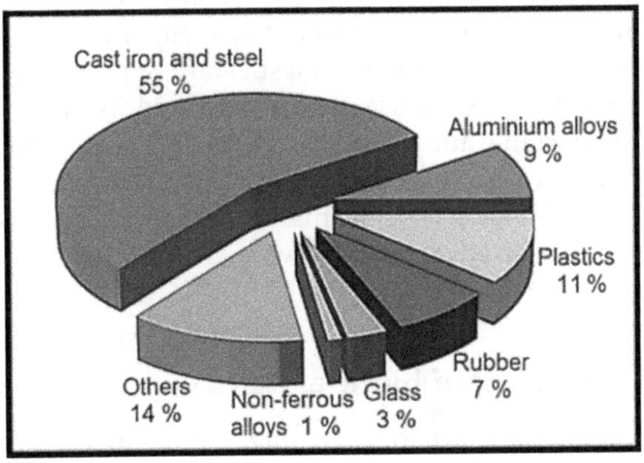

Figure 5.7 Various materials used in the manufacturing of automotive parts [22].

5.2.1 Review of automotive material

The majority (55%) of the elements that make up a vehicle are composed of cast iron, which is the most prevalent element overall. Aluminium alloys are at 14%, with plastics coming in second at 11%. The remaining 14% is made up of different substances as shown in Figure 5.7.

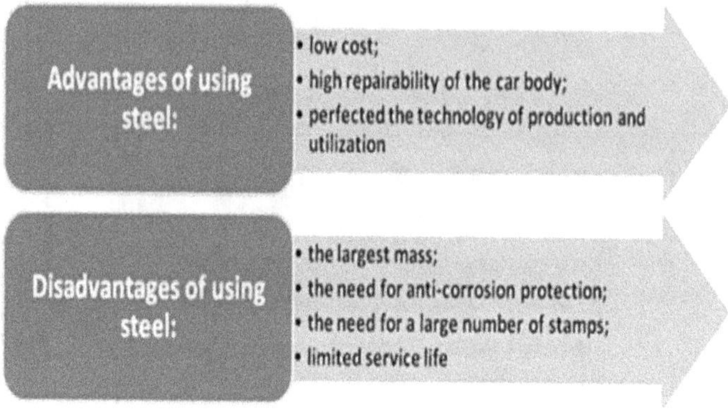

Figure 5.8 Steel's limited use in the automotive sector [22].

Cast iron is the most typical material utilized in automobiles, followed by plastics and aluminium alloys, according to the pie chart. The remaining 14% is made up of different components like rubber, steel, and glass [20, 21]. A variety of elements, such as the material's weight, strength, or cost, influence the way various metals are used in automobiles. Cast iron is a well-liked material for numerous automotive components since it is sturdy, long-lasting, and reasonably priced. Plastics are appropriate for usage in a range of applications, such as interior trim and external body panels, because they are portable and adaptable. Because they are robust and lightweight, aluminium alloys are an excellent option for usage in engine blocks and other elements. While steel is the material that makes up the majority of autos, there are some disadvantages likewise. These limitations are outlined in Figure 5.8.

5.2.2 CAD modeling of bumper

Utilizing PTC's Creo software for designing, a generative 3D modeling programme, the front bumper model was created. By using variables that may be altered to alter the model, parametric modeling enables users to build systems that are determined by those variables. This makes parametric modeling a flexible and efficient way to create models. Creo design software also includes sketching and extrude tools, which are used to create the basic shapes of the model. The sketching tool allows the user to create 2D sketches, which can then be extruded to create 3D solids as shown in Figures 5.9 and 5.10.

The impactor design of the leg form is developed using the extrude product and shell tool. Legform's various components include the 'tibia,

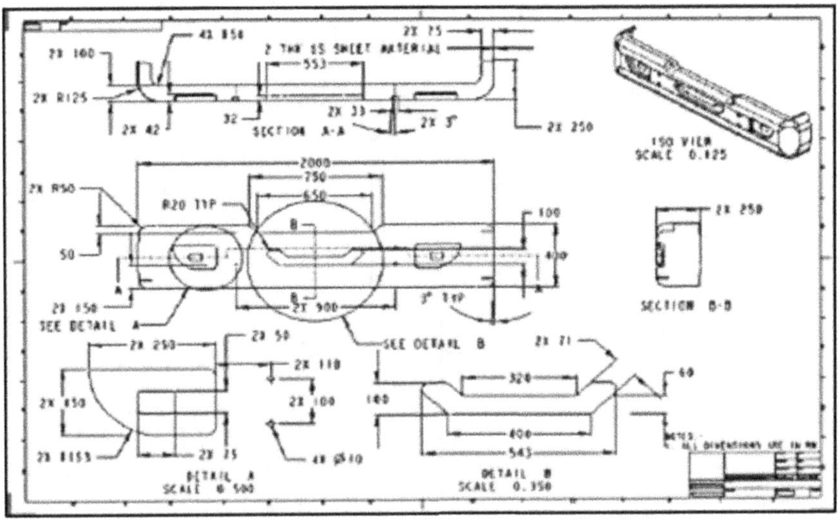

Figure 5.9 Front bumper dimensions.

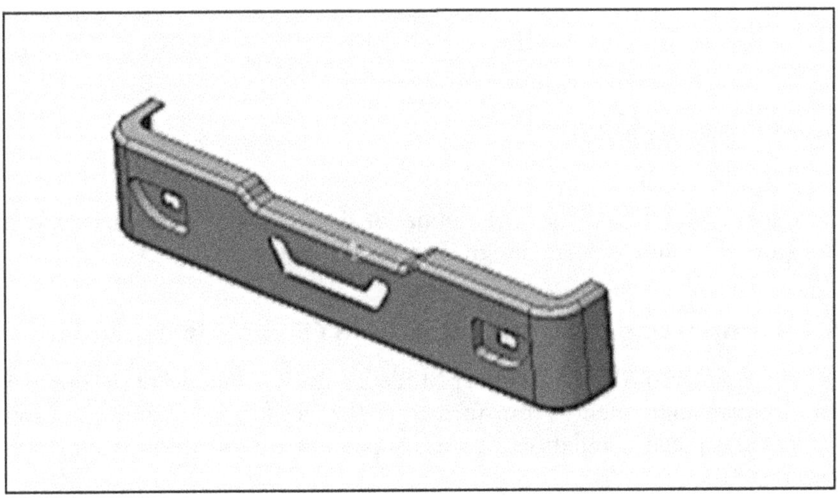

Figure 5.10 Mirroring of extruded feature.

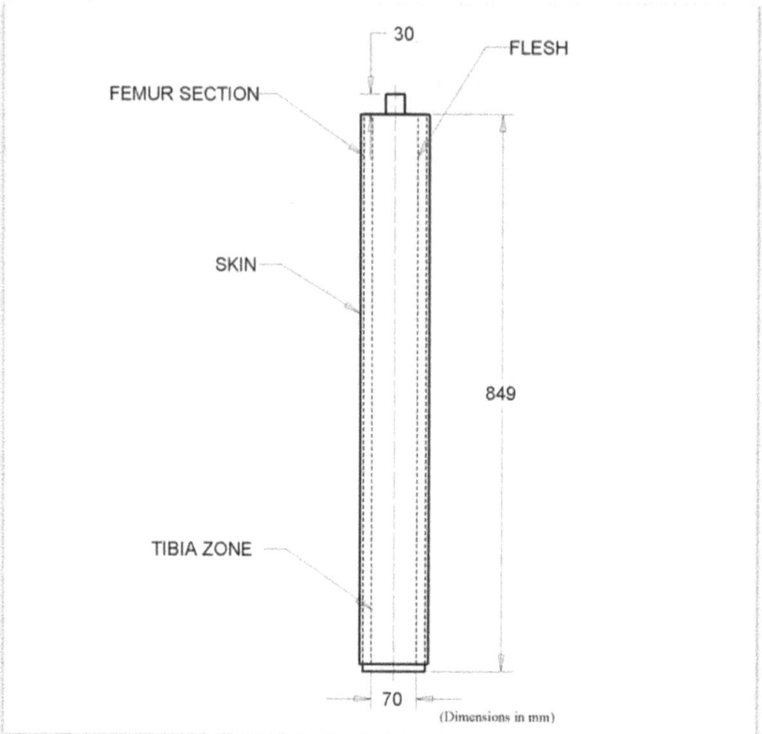

Figure 5.11 Impactor (legform).

skin, foam, and flesh'. The CAD model of the developed bumper is shown in Figure 5.11 and legform design is shown in Figure 5.12.

5.2.3 Importing CAD model in ANSYS explicit dynamics

When loaded into ANSYS Designer Modeller, the CAD model of the bumper and legform component is transformed to the.iges format. Geometric flaws such as sharp edges, inflation, or surface patches are tested for as shown in Figure 5.13.

5.2.4 Meshing CAD model

Bumper and legform meshes are created using ANSYS with fine relevance settings and high transition ratios.

From meshing the number of elements generated is 10032 as shown in Figure 5.14.

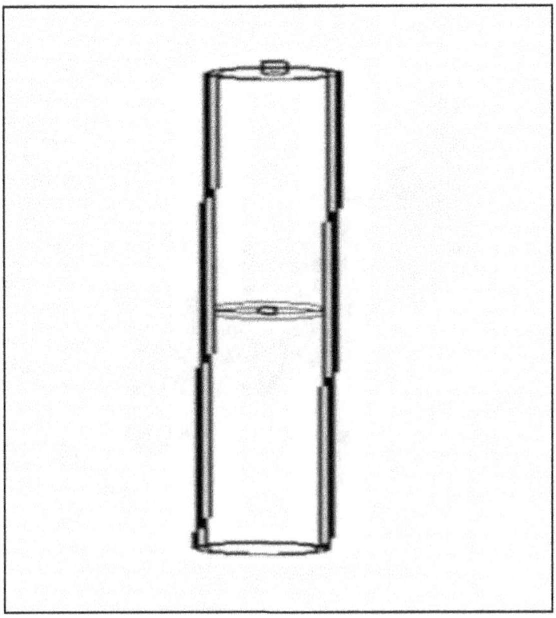

Figure 5.12 EECV impactor CAD model.

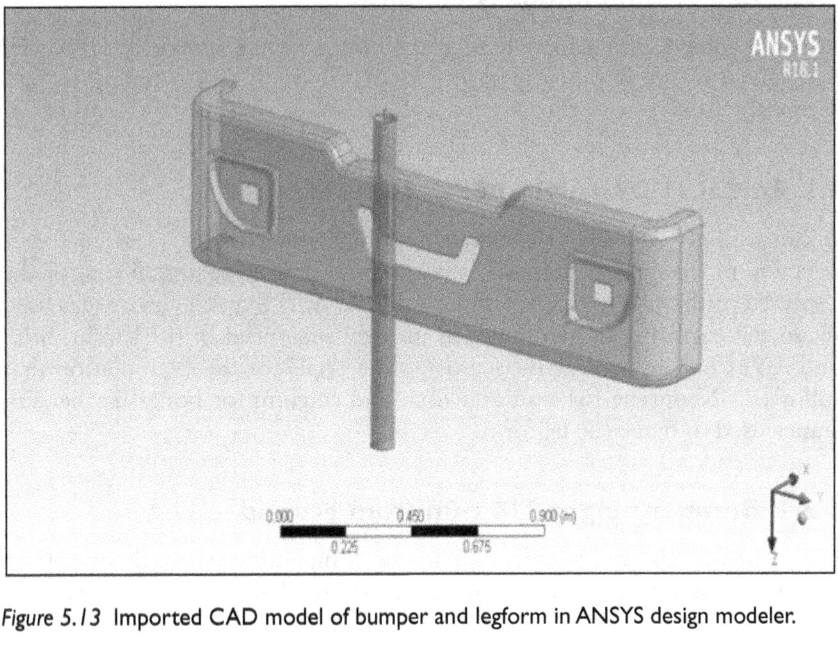

Figure 5.13 Imported CAD model of bumper and legform in ANSYS design modeler.

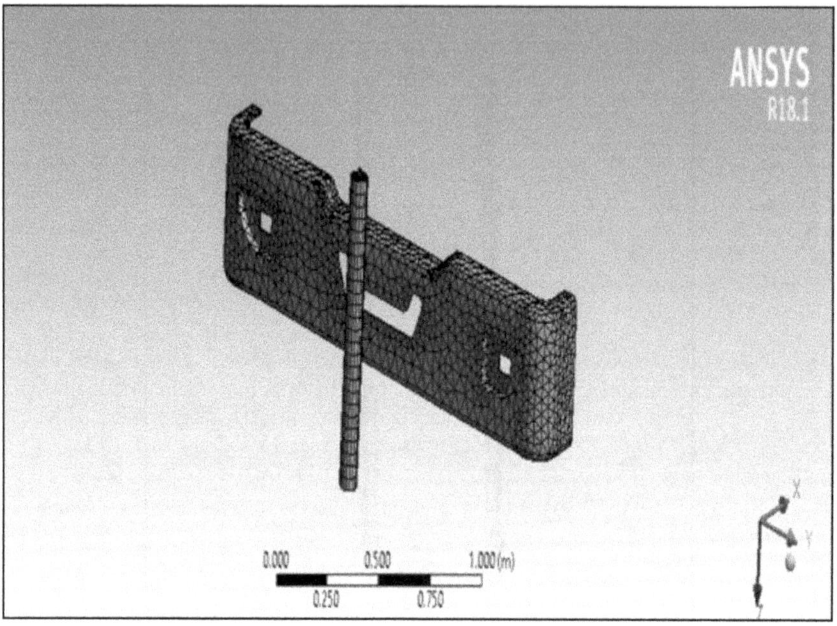

Figure 5.14 Front bumper and legform meshed model.

5.2.5 Loads and boundary condition

Legform is used with a fixed base at the bottom while a velocity of 40 km/hr is given to the bumper as shown in Figure 5.15. Crash timing is .025 seconds.

5.3 EXPLICIT DYNAMIC RESULTS

It can lead to damage of the items, storage problem, or even customer rejection of the ordered items. In such cases, the transportation cost or the expected profit may not be a definite value ANSYS Explicit software is used to do the explicit dynamic analysis at a testing speed of 40 Km/hr. Steel and Al921T were used as the legform's materials for the examination that followed. 'Neoprene for skin and flesh and calcium for bones' is the substance used to make the legform.

5.3.1 Bumper height 210 mm from ground

For a site at 210 mm bumper height, the comparable stress, deformation, and strain energy are assessed. Figure 5.16 shows that the femur bone region

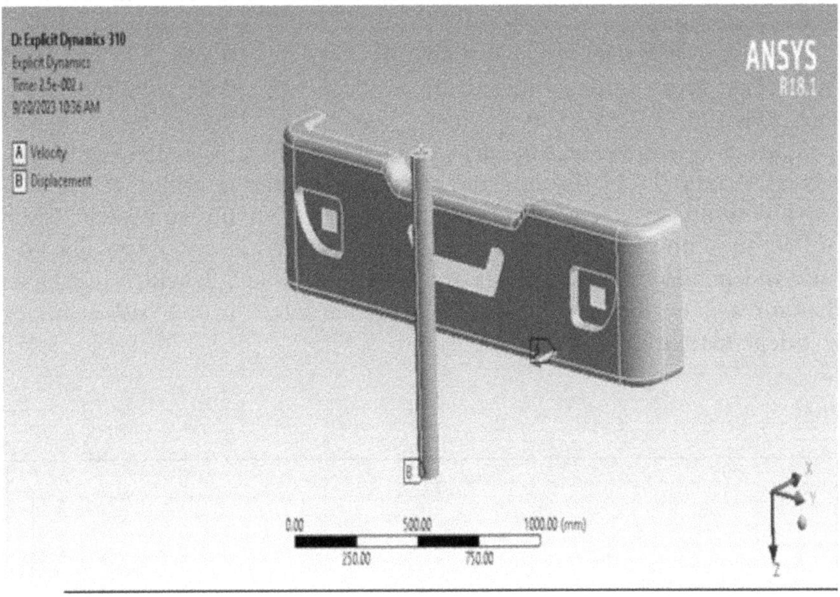

Figure 5.15 Loads and boundary conditions for analysis.

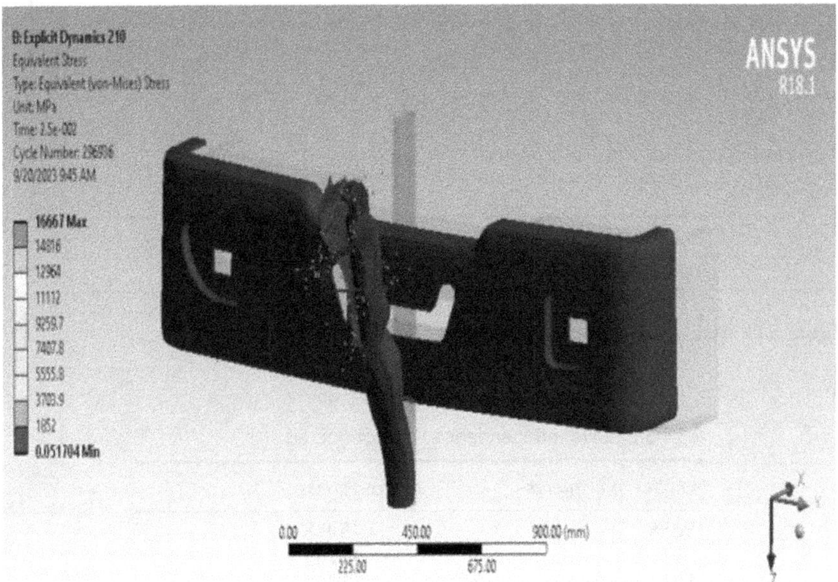

Figure 5.16 Stress for 210 mm height of bumper placement.

has the maximum stress of 1.26 MPa. This maximum stress occurs at 0.025 seconds after impact.

The outer flesh region has a minimum stress of between 0.45 and 0.58 MPa. The stress value obtained are in close agreement with the literature [14].

Figure 5.17 shows the time-dependent stress curve. The stress gradually increases after 0.014 seconds and reaches a maximum value of 1.67e10 Pa. The topmost zone of the legform has the maximum equivalent strain of 0.91 mm/mm, while the bottom zone has the minimum strain. The GIT (grid independence test) findings are shown in Table 5.2, which highlights the thorough evaluation that was done to guarantee the analysis's accuracy and dependability.

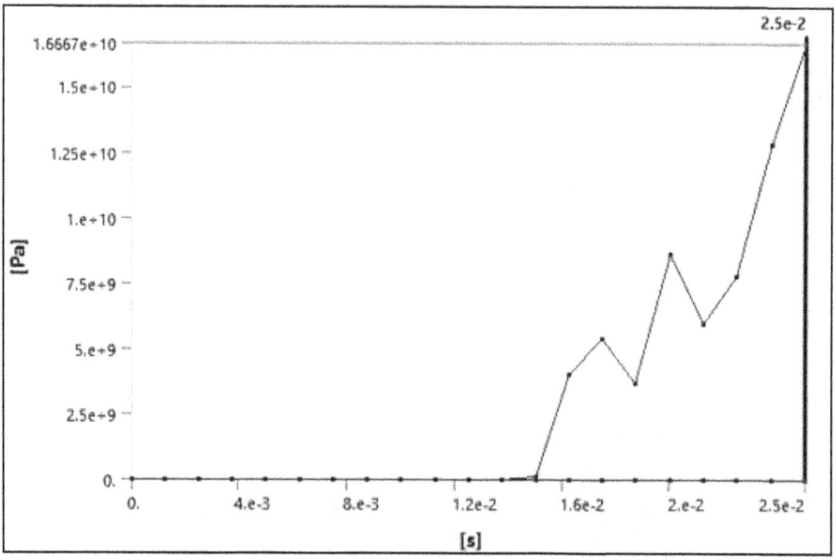

Figure 5.17 Stress versus time for 210 mm height of bumper placement.

Table 5.2 Grid independence test conducted

Number of elements	*Induced stress (MPa)*
9856	15458
9911	15759
9989	15896
10028	16666
10032	16667

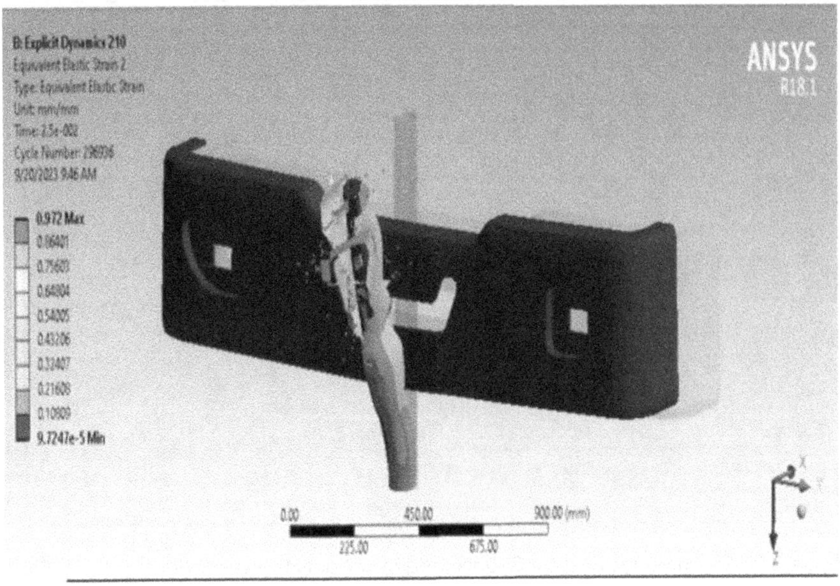

Figure 5.18 Strain for 210 mm height of bumper placement.

The equivalent strain curve shows that the strain gradually increases to a maximum value of 1.14 at 0.025 seconds, and then decreases as shown in Figures 5.18 and 5.19.

The energy absorbed by the legform is shown in Figure 5.20. The curve shows a sudden increase in energy absorption at the time of impact, reaching a maximum value of 139.6 kJ at 0.025 seconds.

The kinetic energy curve shows how the kinetic energy of the legform changes over time as shown in Figure 5.21. The kinetic energy is highest at 0.018 seconds, with a maximum magnitude of 6260 J. The kinetic energy then decreases and then fluctuates as seen by the variation in the curve. The maximum kinetic energy of the legform is 9.1 kJ.

5.3.2 Bumper height 260 mm from ground

For a site at a 260 mm bumper height, the corresponding stress, deformation, and strain energy are analyzed.

Figure 5.22 shows that the femur bone region has the maximum stress of 4.26 MPa. This maximum stress occurs at 0.025 seconds after impact. The outer flesh region has a minimum stress of between 0.9 and 1.405 MPa. Figure 5.23 shows the time-dependent stress curve. The stress gradually increases after 0.014 seconds and reaches a maximum value of 4.35e9 Pa.

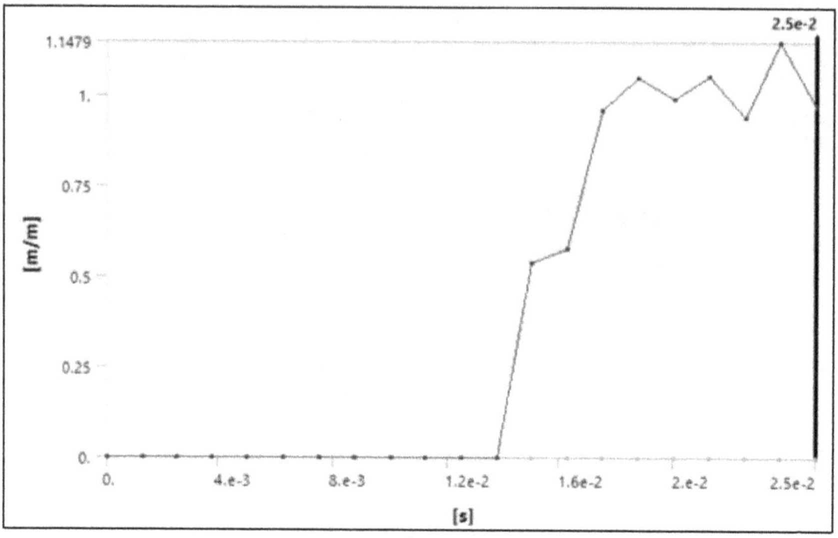

Figure 5.19 Strain versus time for 210 mm height of bumper placement.

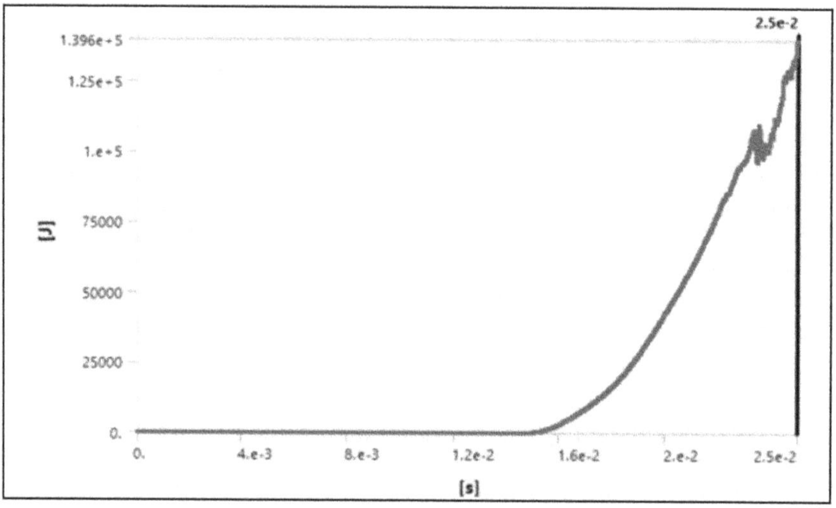

Figure 5.20 Internal energy curve for 210 mm bumper height.

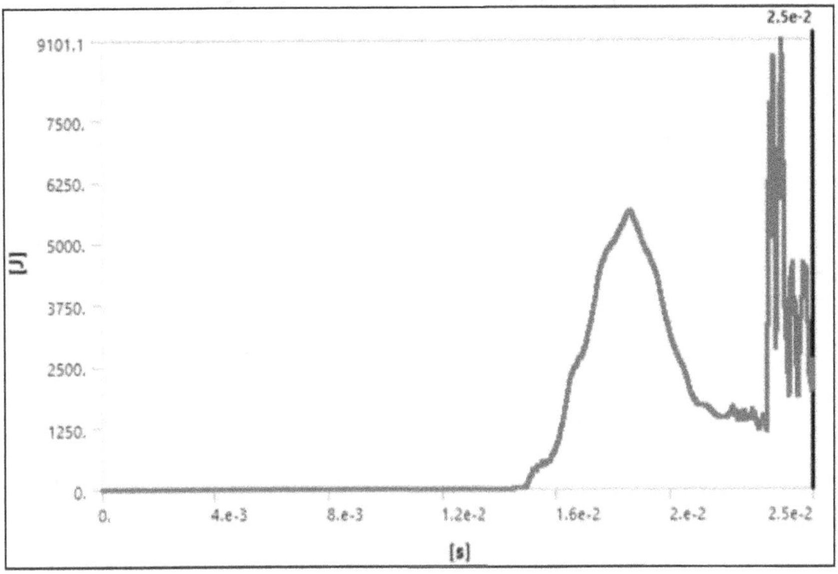

Figure 5.21 Kinetic energy curve for 210 mm bumper height.

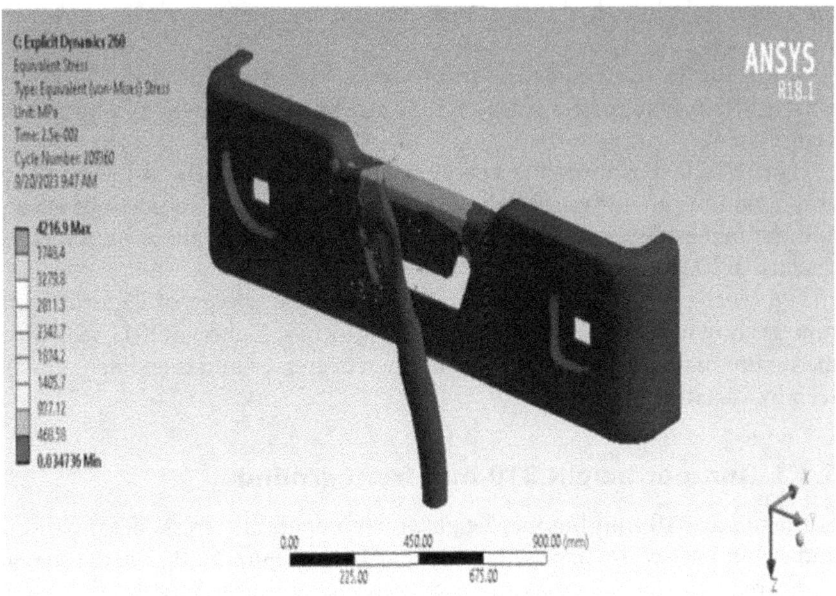

Figure 5.22 Stress for 260 mm height of bumper placement.

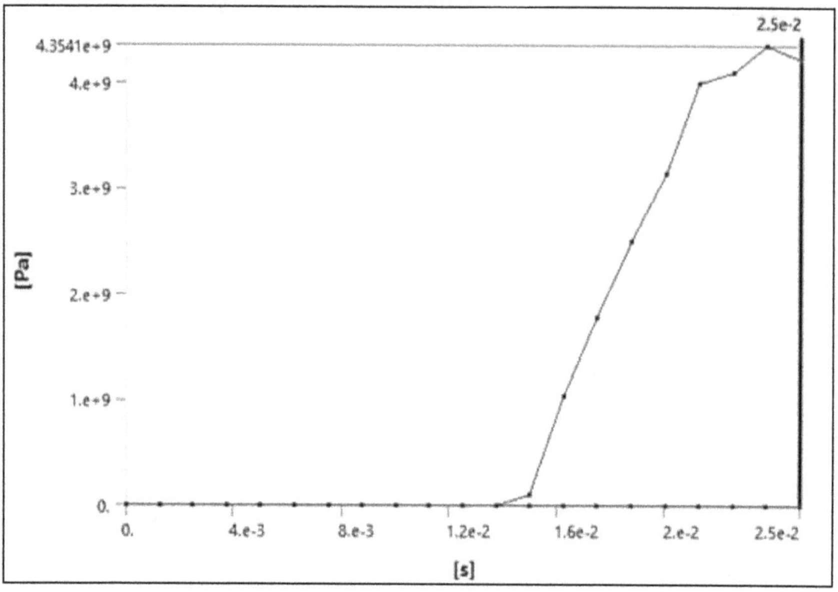

Figure 5.23 Stress versus time for 260 mm height of bumper placement.

The topmost zone of the legform has the maximum equivalent strain of 0.8 mm/mm, while the bottom zone has the minimum strain as shown in Figure 5.24.

According to the equivalent strain curve, the strain progressively rises until it attains its greatest value of 1.03 at.018 before falling as depicted in Figure 5.25.

Figure 5.26 illustrates the energy that the legform has absorbed. The graph exhibits an abrupt rise in energy absorption at the moment of effect, and the highest level of internal energy, with a value of about 46.53 kJ, is reached at .025 secs.

The kinetic energy curve shows varying kinetic energy of legform with time as shown in Figure 5.27. The KE of legform is highest at .017 secs with maximum magnitude of 3.164 kJ and then decreases and then fluctuates as seen by variation in curve.

5.3.3 Bumper height 310 mm from ground

For a site at 310 mm bumper height, the comparable stress, deformation, and strain energy are assessed. It is clear from Figure 5.28 that the femur bone area experiences a maximum stress of magnitude 4.81MPa. The outer flesh area displays minimal stress in the range of .53 to .64 MPa, whereas this maximum stress is found at .025 s after input.

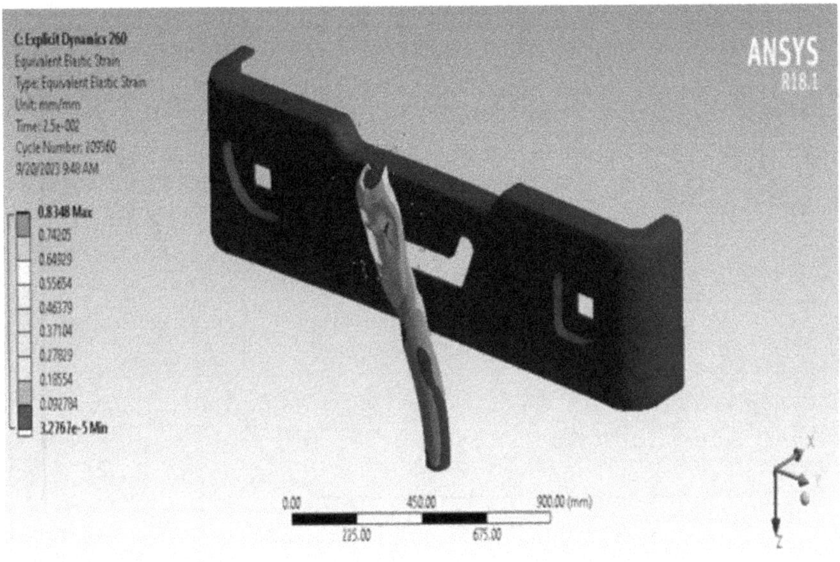

Figure 5.24 Strain for 260 mm height of bumper placement.

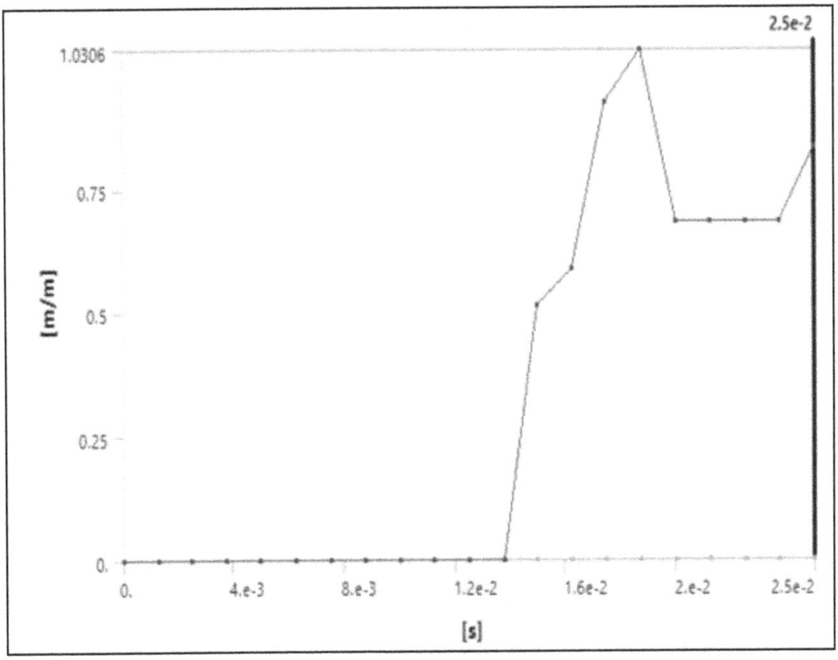

Figure 5.25 Elastic strain curve for 260 mm bumper height.

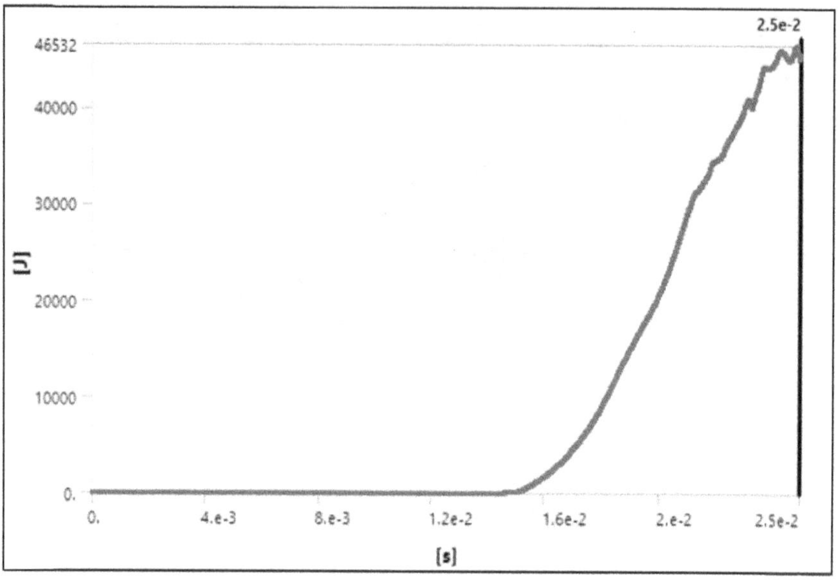

Figure 5.26 Internal energy curve for 260 mm bumper height.

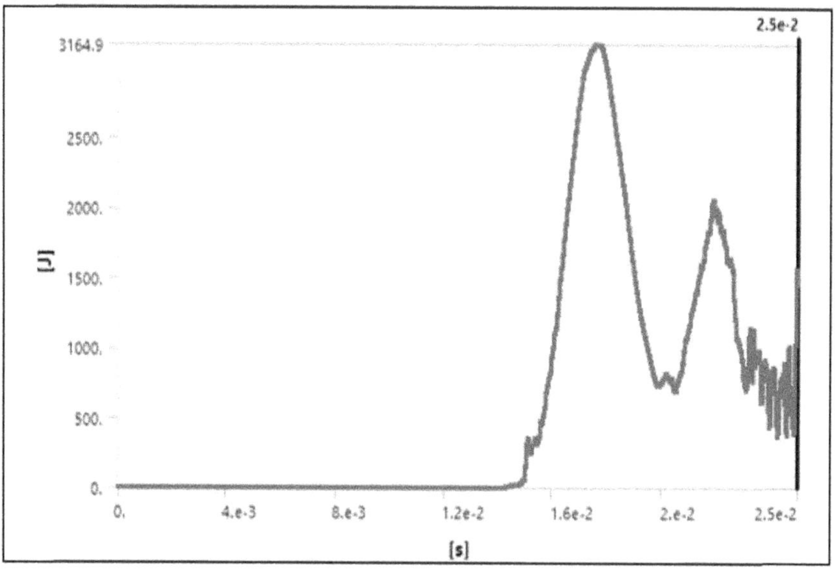

Figure 5.27 Kinetic energy curve for 260 mm bumper height.

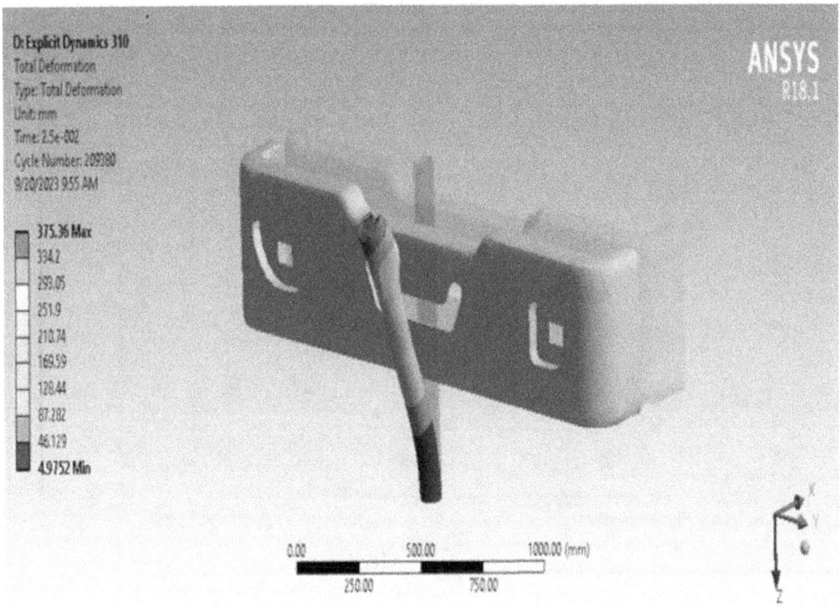

Figure 5.28 Stress for 310 mm height of bumper placement.

Figure 5.29 time-dependent stress curve demonstrates a progressive rise in stress after .014 seconds and a maximum value at 4.81e9Pa. The legform's uppermost zone has an average strain of .63 mm/mm identical, whereas the bottom zone has a minimum strain as shown in Figure 5.30.

According to the equivalent strain curve, the strain progressively rises until it attains a maximum value of .92 at .025 before declining as shown in Figure 5.31.

Figure 5.32 illustrates the energy that legform has received. The graph exhibits a sharp rise in energy absorption at the moment of impact, and the greatest amount of internal energy, with a magnitude of about 53.12 kJ, is reached at .025 secs.

The legform's kinetic energy changes over time as shown by the kinetic energy graph. The legform's KE is at its greatest at 0.018 secs with a maximum value of 2.16 J, but it then starts to decline and change as the curve changes as shown in Figure 5.33. When an object collides, its acceleration is very strong. The curve exhibits both a quick spike and drop in acceleration. Table 5.3 is a comparing matrix of kinetic energy, internal energy, and comparable elastic strain; the following figures offer more in-depth explanations.

As shown in Figure 5.34, the equivalent strain in the legform impactor model decreases with increasing bumper height. The minimum equivalent

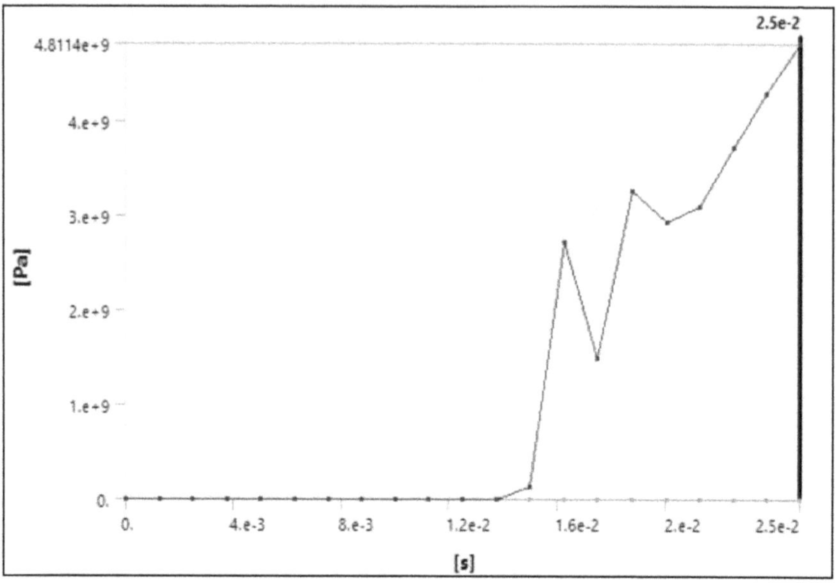

Figure 5.29 Stress versus time for 310 mm height of bumper placement.

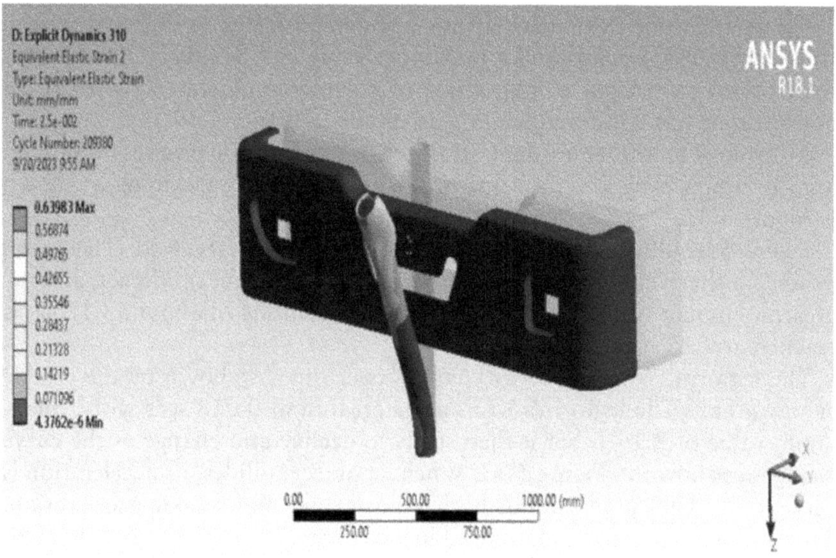

Figure 5.30 Strain for 310 mm height of bumper placement.

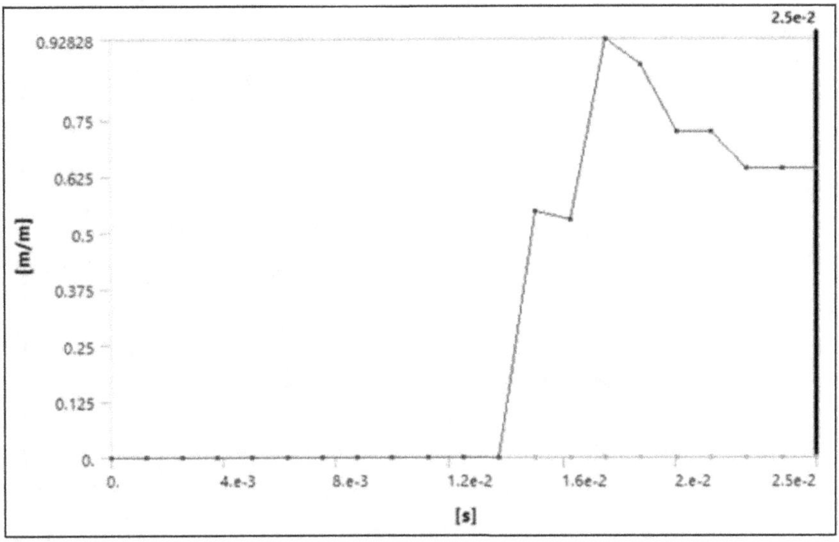

Figure 5.31 Strain versus time for 310 mm height of bumper placement.

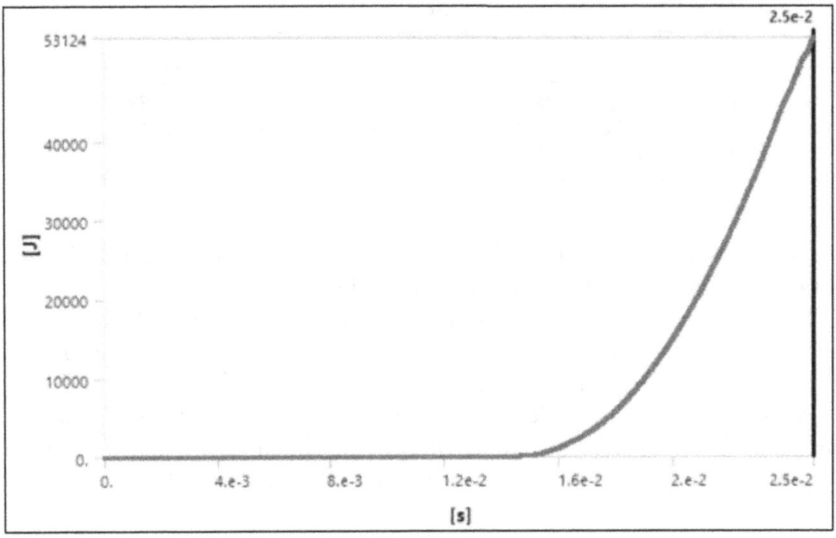

Figure 5.32 Internal energy curve for 310 mm bumper height.

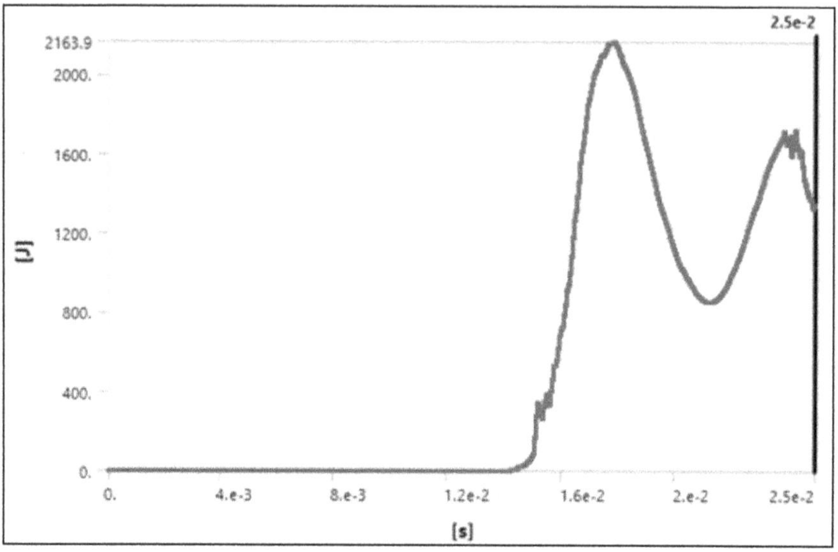

Figure 5.33 Kinetic energy curve for 310 mm bumper height.

Table 5.3 Comparison matrix table

Height	*210 mm*	*260 mm*	*310 mm*
Equivalent elastic strain (m/m)	1.1479	1.0306	.928
Max. internal energy (kJ)	139.6	16.53	53.12
Max. kinetic energy (kJ)	9.101	3.1641	2.163

strain is observed for 310 mm bumper height, and the maximum equivalent strain is observed for 210 mm bumper height.

Figure 5.35 displays the internal energy comparison plot. The comparative plot demonstrates that at bumper heights of 210 mm and 260 mm, respectively, internal energy is at its greatest or least values.

As bumper height rises, kinetic energy in the legform decreases as shown in Figure 5.36. Maximum kinetic energy is attained for bumper heights of 210 mm, while minima are noted for bumper heights of 310 mm.

5.4 CONCLUSION

The bumper placement has a significant effect on the nature and type of injury caused to pedestrians. Explicit dynamic analysis can be used to

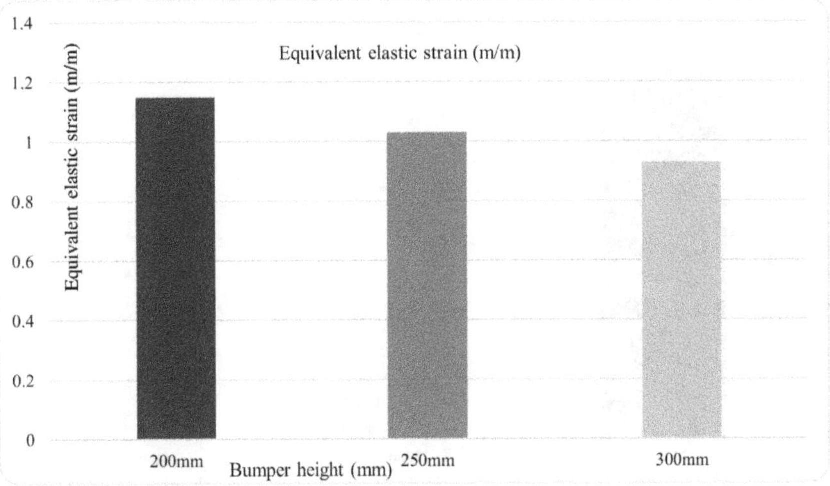

Figure 5.34 Equivalent elastic strain versus bumper height.

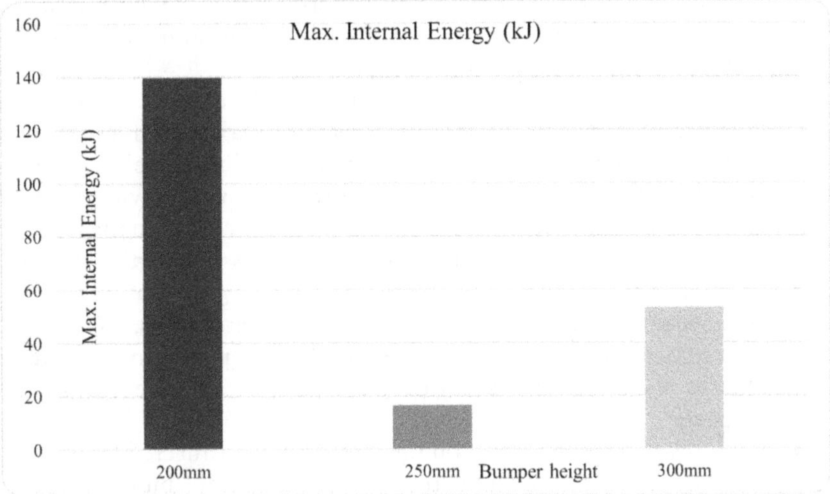

Figure 5.35 Maximum internal energy versus bumper height.

identify the possible regions of high injury to pedestrians by analyzing internal energy, shear stress, and energy absorption. Explicit dynamics has significantly reduced the time and cost involved in testing. The crash analysis results at 40 km/h are as follows:

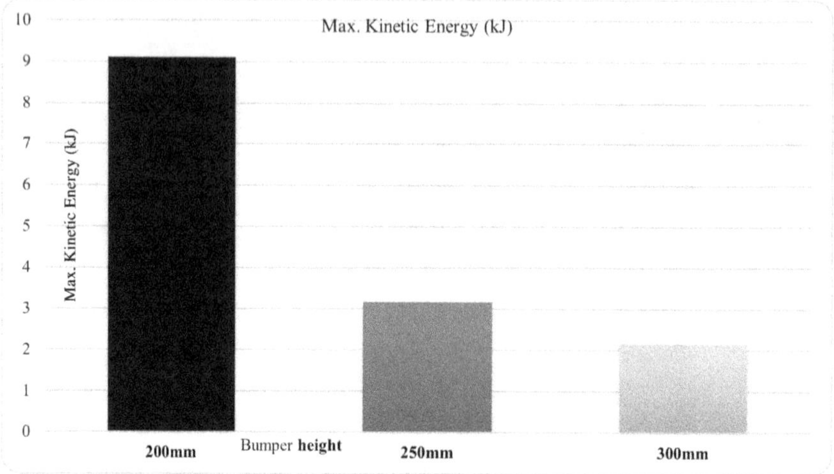

Figure 5.36 Kinetic energy versus bumper height.

- The strain generated in the EECV legform decreases with increasing bumper placement height, with the minimum strain observed at 310 mm bumper placement.
- The internal energy generated in the EECV legform varies with bumper placement. The internal energy is maximum for 210 mm bumper placement and minimum for 260 mm bumper placement.
- The impulse generated on the legform also varies with bumper placement height. The maximum impulse is generated on the legform for 210 mm bumper placement, which decreases with further increase in bumper placement.
- The location of injury in the legform/pedestrian changes with time. The maximum damage is observed on the lower portion of the tibia due to shearing. This shearing may result in fracture due to the impact.
- The shear stress generated on the legform is also affected by bumper height. The shear stress increases with decreasing bumper height. The damage to the pedestrian can be evaluated through the shear stress generated. The 310 mm bumper height appears to have the smallest shear stress.
- The research conducted would enable manufacturers to develop bumpers which are more effective in reducing pedestrian injuries.
- The study's findings demonstrated that the pedestrian injury was significantly influenced by the bumper height. The severity of the pedestrian injury increases with decreasing bumper height. The research

also demonstrated that Al921T's superior energy absorption capabilities over the original bumper material led to less severe injuries for pedestrians.

In order to increase the accuracy of the findings, additional research should look into the following issues:

- Research into using radar systems to autonomously identify pedestrians and items in a vehicle's route. Radar systems may be used to autonomously identify pedestrians or objects in a vehicle's route. The radar sensor could notify the vehicle's breaks to the presence of an item or pedestrian so that they are able to apply the brakes to prevent a collision.
- Examining the viability of airbags for hoods and bumpers: in the case of an accident, airbags could be utilized to lessen passenger casualties. In order to prevent the passenger from being hit by the object or pedestrian which the car collides with, airbags might be put in the hood or bumper of the automobiles.
- Energy-absorbing components could be utilized to lessen the force of contact in a collision. This is why energy-absorbing front-end components or bonnets have been put through small-velocity collision testing. To shield the occupants from harm, these materials might be utilized on the front end of a car, including the bumper and bonnet.
- Further research can be conducted on improving design of bumpers using techniques of additive manufacturing.
- The design of the front bumper can be optimized using the response surface optimization technique and genetic algorithm techniques.

REFERENCES

[1] *Traffic Safety Facts*. "A compilation of motor vehicle crash data from the fatality analysis reporting system and the general estimates system," Report No. DOT HS 809 (2005): 631.

[2] T.-H. Lee, G.-H. Yoon, M.-S. Han, and S.-B. Choi, "Shock mitigation of pedestrians from sports utility vehicles impact using active pop-up and extended hood mechanisms: experimental work," *Proceedings of the Institution of Mechanical Engineers, Part D: Journal of Automobile Engineering*, vol. 232, no. 12, pp. 1573–1583, Oct. 2018, doi: 10.1177/0954407017732641.

[3] B. Busov and M. Bartlova, "Improvement of active hinge of the car bonnet," in *Proceedings of the 16th International Conference on Mechatronics – Mechatronika 2014*, IEEE, Dec. 2014, pp. 575–581. doi: 10.1109/MECHATRONIKA.2014.7018323.

[4] J. Li, L. Yao, X. Xu, B. Cheng, and J. Ren, "Deep reinforcement learning for pedestrian collision avoidance and human-machine cooperative driving," *Information Sciences (N Y)*, vol. 532, pp. 110–124, Sep. 2020, doi: 10.1016/j.ins.2020.03.105.

[5] Sudirja, A. Hapid, S. Kaleg, A. C. Budiman, and Amin, "The crumple zone quality enhancement of electric cars bumper fascia using a carbon fiber reinforced vinyl Ester – microsphere composites," in *2019 International Conference on Sustainable Energy Engineering and Application (ICSEEA)*, IEEE, Oct. 2019, pp. 1–5. doi: 10.1109/ICSEEA47812.2019.8938633.

[6] M. Baek, D. Jeong, D. Choi, and S. Lee, "Vehicle trajectory prediction and collision warning via fusion of multisensors and wireless vehicular communications," *Sensors*, vol. 20, no. 1, p. 288, Jan. 2020, doi: 10.3390/s20010288.

[7] X. Zhang, Z. Xue, and W. Tu, "Design and performance research of a wearable airbag for the human body," *Applied Sciences*, vol. 13, no. 6, p. 3628, Mar. 2023, doi: 10.3390/app13063628.

[8] A. A. Senthilnathan, S. Rajendra Mehta, and G. Kabir, "Bumper beam composite material selection using fuzzy multi-criteria analysis," in *2022 International Conference on Decision Aid Sciences and Applications (DASA)*, IEEE, Mar. 2022, pp. 243–247. doi: 10.1109/DASA54658.2022.9765059.

[9] A. Th. Schulz and R. Stiefelhagen, "Pedestrian intention recognition using Latent-dynamic conditional random fields," in *2015 IEEE Intelligent Vehicles Symposium (IV)*, IEEE, Jun. 2015, pp. 622–627. doi: 10.1109/IVS.2015.7225754.

[10] C. Luo, G. Xin, H. Xu, and W. Tang, "Glare-free high-beam control for oncoming vehicle safety in nighttime," *IEEE Consumer Electronics Magazine*, vol. 10, no. 2, pp. 8–15, Mar. 2021, doi: 10.1109/MCE.2020.3040724.

[11] D. Singh and C. K. Mohan, "Deep spatio-temporal representation for detection of road accidents using stacked autoencoder," *IEEE Transactions on Intelligent Transportation Systems*, vol. 20, no. 3, pp. 879–887, Mar. 2019, doi: 10.1109/TITS.2018.2835308.

[12] G. Lanzaro, C. Fu, and T. Sayed, "Comparing the crash risk of vehicle-pedestrian interactions using autonomous vehicle data," in *2023 IEEE Intelligent Vehicles Symposium (IV)*, IEEE, Jun. 2023, pp. 1–6. doi: 10.1109/IV55152.2023.10186754.

[13] V. Akshitha Sumahi, S. P. Jani, and S. Uppalapati, "Impact study of a car bumper by using carbon fiber reinforced polyetherimid and S-glass/epoxy composite," *Mater Today Proc*, May 2023, doi: 10.1016/j.matpr.2023.05.177.

[14] M. A. Basith, N. C. Reddy, S. Uppalapati, and S. P. Jani, "Crash analysis of a passenger car bumper assembly to improve design for impact test," *Mater Today Proc*, vol. 45, pp. 1684–1690, 2021, doi: 10.1016/j.matpr.2020.08.561.

[15] L. Kumar, A. Kumar, R. K. Sharma, and P. Kumar, "Smart manufacturing and industry 4.0," in *Handbook of Smart Manufacturing*, Boca Raton: CRC Press, 2023, pp. 1–28, doi: 10.1201/9781003333760-1.

[16] A. Kumar, H. Singh, P. Kumar, and B. AlMangour, *Handbook of Smart Manufacturing,* Boca Raton: CRC Press, 2023, doi: 10.1201/9781003333760.

[17] Ajay Kumar, Parveen Kumar, Ashish Kumar Srivastava, and Vikas Goyat, Modeling, characterization, and processing of smart materials. in *Advances in Chemical and Materials Engineering*, IGI Global, 2023, doi: 10.4018/978-1-6684-9224-6.

[18] A. Kumar, P. Kumar, H. Singh, A. Haleem, and R. K. Mittal, "Integration of reverse engineering with additive manufacturing," in *Advances in Additive Manufacturing Artificial Intelligence, Nature-Inspired, and Biomanufacturing*, Elsevier, 2023, pp. 43–65, doi: 10.1016/B978-0-323-91834-3.00028-4.

[19] A. Agarwal and L. Mthembu, "Design and response surface optimization of heavy motor vehicle chassis using P100/6061 Al MMC," in *Recent Advances in Materials and Modern Manufacturing*, Lecture No., D. P. I. A. Palani, P. Sathya, Ed., Singapore: Springer Singapore, 2022, pp. 1–12, doi: 10.1007/978-981-19-0244-4_1.

[20] A. Agarwal and L. Mthembu, "Structural analysis and optimization of heavy vehicle chassis using aluminium P100/6061 Al and Al GA 7-230 MMC," *Processes*, vol. 10, no. 2, p. 320, Feb. 2022, doi: 10.3390/pr10020320.

[21] A. Agarwal and L. Mthembu, "Modelling and FE simulation of HVC using multi objective response surface optimization techniques," *Revue des composites et des matériaux avancés*, vol. 31, no. 6, pp. 307–315, Dec. 2021, doi: 10.18280/rcma.310601.

[22] T. P. Hovorun, K. V. Berladir, V. I. Pererva, S. G. Rudenko, and A. I. Martynov, "Modern materials for automotive industry," *Journal of Engineering Sciences*, vol. 4, no. 2, pp. f8–f18, 2017, doi: 10.21272/jes.2017.4(2).f8.

Chapter 6

Improvise in sigma level in automobile industries

Aman Kumar and Sachin Kumar
Department of Mechanical Engineering, UIET, MDU, Rohtak, India

6.1 INTRODUCTION

In this modern era, profitability is on the rise. Everyone is planning to enhance their methods, eliminate access efforts, minimize waste, smooth processes, remove variation, etc. Industrialists are using several techniques and methodologies to make their process variations waste-free. It also contributes to our ecosystem, industry, and organization. Six-sigma is a statistical tool that is used to improve processes, eliminate wastage, reduce costs, etc. Six-sigma methodology is used in the DMAIC technique based on the PDCA cycle, which implies plan-do-control-act [1–4]. It has its own strengths and weaknesses, and there is no right or wrong with this technique. Now, it has become the statistical tool or symbol generally used in mathematics to determine the standard deviation. It is a technical or scientific method or process used for process enhancement or improvement planning, and the method used for process improvement is generally known as DMAIC [5–6]. The target is 3.4 errors per minute by using the technique DMAIC, which is a process improvement model. Six sigma's methodology has main focuses on high productivity, quality improvement, waste elimination, and defect reduction; it also lays down the flaw rate target at three point four errors per million opportunities [7–9]. It results in superior, fast, and cost-efficient products. Six-sigma is an improvement approach that focuses on client requirement defects, cycle time prevention, reduction, and cost savings [10–13].

6.2 PROBLEM FORMULATION

The present study deals with the analysis of the gap between the knob and lever of a door handle in a car using the DMAIC technique to obtain the desired quality. The car door handle is an essential part that helps open the door from the inside. Small defects in the door handle may block

 DOI: 10.1201/9781003502470-6

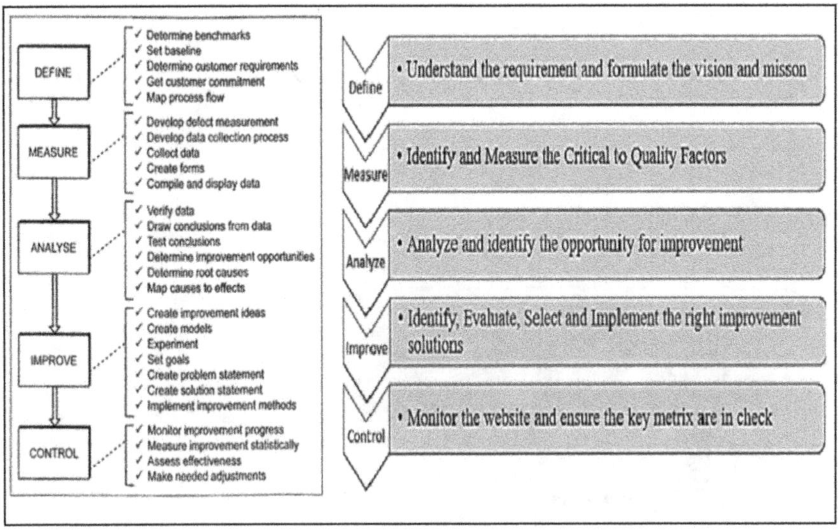

Figure 6.1 DMAIC elaboration.

exits from the car (Figure 6.1). It comes under the 'B' category component, which belongs to a category in which an accident may take place and the functioning of the car will fail but there will be no loss of human life involved, whereas category 'A' belongs to a category where loss of human life involves, and if there is no failure and only aesthetics reduce the charm of the parts, then this falls under category 'C'. In the present study, DMAIC methodology is applied to resolve the problem (Figure 6.1), which consists of the following phases: 1. Define, 2. Measure, 3. Analysis, 4. Improve, 5. Conclusion.

6.3 DMAIC ANALYSIS

6.3.1 Define

Currently, the gap between the knob and lever is 2 mm, which is very far from the permissible tolerance. In this defined phase, a **SIPOC** (supplier-input-process-output-customer) process was made for handling the gap (Figure 6.2). The current rejection rate is too high due to the gap between the knob and lever. The main aim of the present work is to reduce the rejection rate of door handles, which was due to the high gap between knob and lever (Figure 6.3).

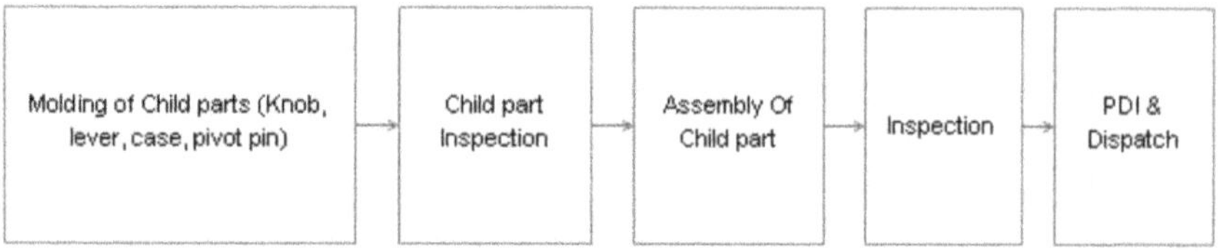

Figure 6.2 SIPOC flow chart.

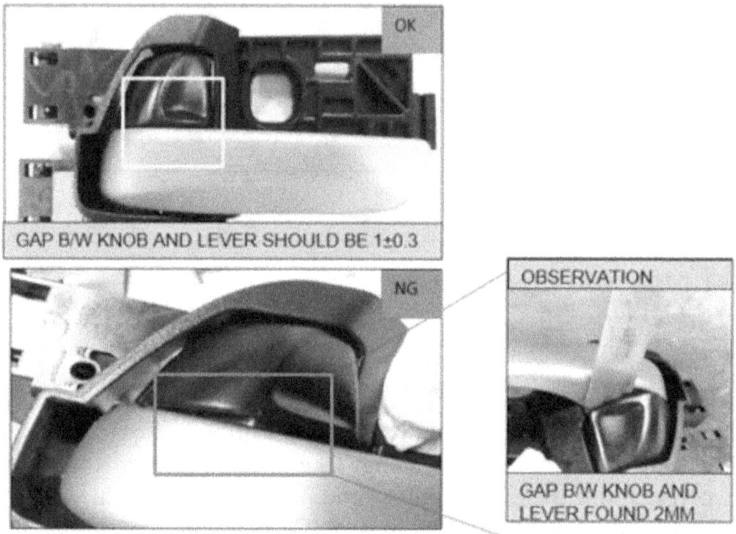

Figure 6.3 Problem occurrence.

6.3.2 Measure

In this step, a system measurement takes place, which includes an R&R gauge study analysis in which if there are less than 10% variations measured in gauge values, then the design is acceptable; if the gauge variations are more than 30%, then the part should be rejected. If the variations are within the marginal range, i.e., 10–30% gauge variations, then improvement steps should be taken to reduce the marginal gauge values. If the gauge variation result is unacceptable, then consider the R&R (repeatability and reproducibility) for improvement. It follows two methods, viz.,

- Iterative testing of composition, also known as repeatability, in which multiple times the same person measures the same variations, where 12.5% of measurement error and 87.5% of the entire variation of process ocuurance were measured (Figure 6.4).

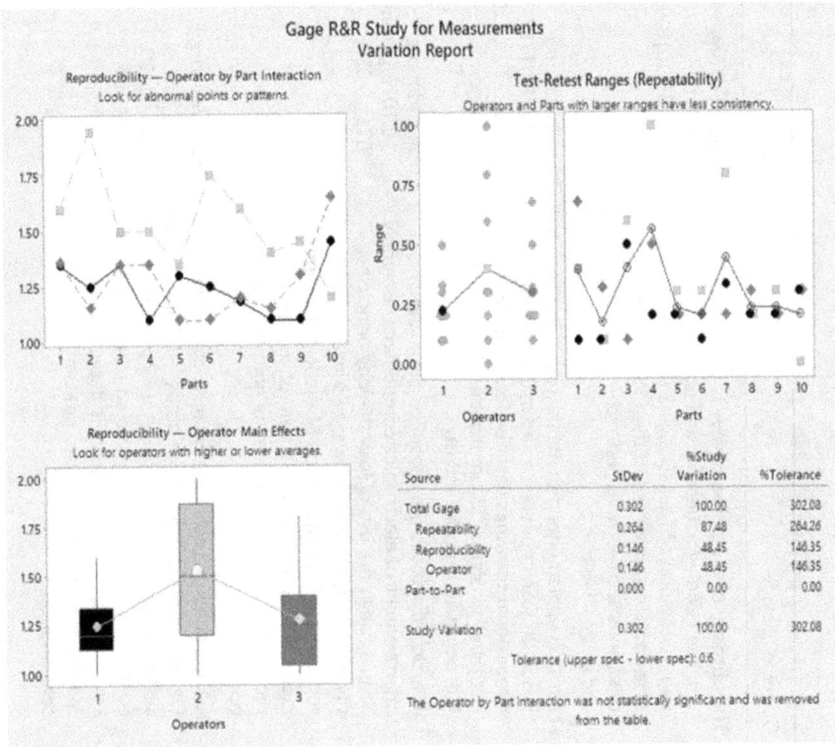

Figure 6.4 R&R study analysis.

- Operational or reproducibility, in which different people measure the same item with variation, with 48.4% measurement error and 51.6% of the entire process variation shown in Table 6.1.

6.3.3 Analysis

During this phase, the project associates examine the data collected in the measurement phase and seek explanation for the reading by using tools like brainstorming and cause-and-effect diagrams. During brainstorming analysis, the root cause was selected by the process of 'know-how' and brainstorming sessions with the respective team members. On the basis of these causes and effects, a cause-and-effect diagram was made. This diagram shows the effect and causes of a fishbone pattern, and it is also known as the fishbone diagram. On further brainstorming and discussion, the cause-and-effect diagram is remade. Minitab 19.0 software is used for process capability analysis for door handles and over-all or actual capability is defined

Table 6.1 Literature outcome in tabulated form

Sr. no	*Author/reference literature*	*Title of chapter*	*Outcome of study*	*DOI*
1	Singh, Mahipal, and Rajeev Rathi	A Structured Review of Lean Six Sigma in Various Industrial Sectors	The purpose of the present study is to expose the detailed review for benefits and challenges about implementation of lean six sigma (LSS) in business organization and spread of the LSS literature in terms of various sectors wise, research methodology wise and journal wise.	https://doi.org/10.1108/ijlss-03-2018-0018.
2	Elboq, Raja, Mustapha Hlyal, and Jamila El Alami	Empirical Assessment of Critical Success Factor of Lean and Six Sigma in the Moroccan Automotive Industry	LM and SS implementing characteristics were revealed. Based on CSF importance and maturity, a bi-dimensional scan that describe lean manufacturing (LM) and six-sigma (SS) implementation within the Moroccan automotive industry.	https://doi.org/10.1088/1757-99x/827/1/012043.
3	Bastos, Nuno Miguel Moreno Teixeira, Anabela Carvalho Alves, Felipe Castro, Júlio César Duarte, LuíS Pinto Ferreira, and F.J.G. Silva	Reconfiguration of Assembly Lines Using Lean Thinking in an Electronics Components	This article discusses the reconfiguration and improvement of electronics components' assembly lines through the use of lean thinking principles. Overall, the level of scrap was reduced by 57% through the implementation of the lean six-sigma methodology, and the cycle time of an operation was reduced by 12,5% by eliminating non-value adding operations.	https://doi.org/10.1016/j.promfg.2021.10.053
4	Yang, Hui, Prahalada Rao, Timothy W. Simpson, Yan Lu, Paul Witherell, Abdalla R. Nassar, Edward W. Reutzel, and Soundar Kumara	Six-Sigma Quality Management of Additive Manufacturing	The six sigma (6S) technique entails a data-driven DMAIC methodology of five steps—defines measure, analyze, improve, and control. A new process of control approaches was discussed to optimize the action plans, once an anomaly is detected, with specific consideration of lead time and energy consumption.	https://doi.org/10.1109/jproc.2020.3034519

5	Gaikwad, Lokpriya, Umesh Bhushi, and S. N. Teli	Implementation of Six Sigma Methodologies to Gain a Competitive Advantage: A Case Study Approach	Six-sigma DMADV (define-measure-analyze-design-verify) methodologies is an influential approach to designing products, processes, and services to fulfill the requirements and potential of the customer while reducing the cost of quality. DMADV uses influential and constructive statistical tools to forecast and enhance quality before prototypes are built.	https://doi.org/10.1109/aset53988.2022.9735103
6	Bewoor, Anand K., Mitali U. Kharul, Saloni G. Gosavi, and Tanvi J. Kuray	Application of 7-Step Problem-Solving Methodology for Defect Elimination: A Case Study in an Automotive Industry	7-step problem-solving methodology (7SPSM), along with 8D, PDCA, DMAIC, and six sigma are adopted. From the results experienced, it can be concluded that 7SPSM provides an effective methodology not only for solving problems but also for continual improvement by elimination of recurrences of product failure.	https://doi.org/10.1007/978-981-16-9952-8_60
7	Kumar, V. Ravi, Vikas Modgil, and Ajay Kumar	Stochastic Petri Nets Modelling for Performance Assessment of a Manufacturing Unit	The Petri Nets modeling method run the plant in virtual manner and the system availability is found to be 80.27% substituting the input parameters in the timed transitions of each sub-system. Further, the performance assessment through Patri Nets Model is compared with the Markov model solution to verify the findings.	https://doi.org/10.1016/j.matpr.2022.01.073
8	Kumar, Ajay, Hari Singh, Parveen Kumar, and Bandar AlMangour	*Handbook of Smart Manufacturing*	All new concepts and innovative appraoches are mentioned in this book.	https://doi.org/10.1201/9781003333760
9	Ajay, Ashwini Kumar, Parveen, Rajesh Kumar Goel, and Ravi Kant Mittal	*Waste Recovery and Management*	Applications of six-sigma technique to resolve the problems of waste management industries.	https://doi.org/10.1201/9781003359784

(*continued*)

Table 6.1 (*Cont.*)

Sr. no	*Author/reference literature*	*Title of chapter*	*Outcome of study*	*DOI*
10	Rani, Sangeeta, Khushboo Tripathi, and Ajay Kumar	Machine Learning Aided Malware Detection for Secure and Smart Manufacturing: A Comprehensive Analysis of the State of the Art.	Implementation of a novel approach for smart manufacturing using malware detection.	https://doi.org/10.1007/s12008-023-01578-0
11	Rani, Sangeeta, Khushboo Tripathi, Yojna Arora, and Ajay Kumar	Analysis of Anomaly Detection of Malware Using KNN	A comprehensive study of anomaly detection of malware based on machine learning algorithms is presented here. This chapter also explains about the implementation of k-nearest neighbors of anomaly detection and discusses the challenges associated with implementing malware classifiers.	https://doi.org/10.1109/iciptm54933.2022.9754044
12	Joshi, Vishwas Deep, Priya Agarwal, and Ajay Kumar	Fuzzy Transportation Planning: A Goal Programming Tactic for Navigating Uncertainty and Multi-Objective Decision Making.	This approach empowers a company to generate compromise solutions aligned with their preferences and illustrated through a practical MOTP solved with LINGO software. This approach effectively removes uncertainty and enhances decision-making by generating compromise solutions aligned with preferences.	https://doi.org/10.1007/s12008-023-01634-9
13	Sathish kumar Arumugam, Sachin Kumar, Pramod Sridhara, Srinivasan Raju, Ashwin Prabhu Gnanasekaran, Nanthakumar Sivasamy and Thangarajan Sivasankaran Senthilkumar	Machine Learning-Based Investigation of Wear and Frictional Behavior in Graphite-Reinforced Aluminum Nanocomposites	The researcher used a cell segmentation technique in conjunction with other image analysis methods to quantitatively retrieve and compute the cellular microstructural structures in a sub-grain size of silicon carbide (SiC)-reinforced AA2219 made by powder fusion bed (size 0.5–1 μm). Over 83 geometric features were retrieved and statistically analyzed using ML (machine learning) techniques to examine the structure-property relationships in SiC-reinforced AlSi20Mg nanocomposites	https://doi.org/10.17756/nwj.2023-s3-054

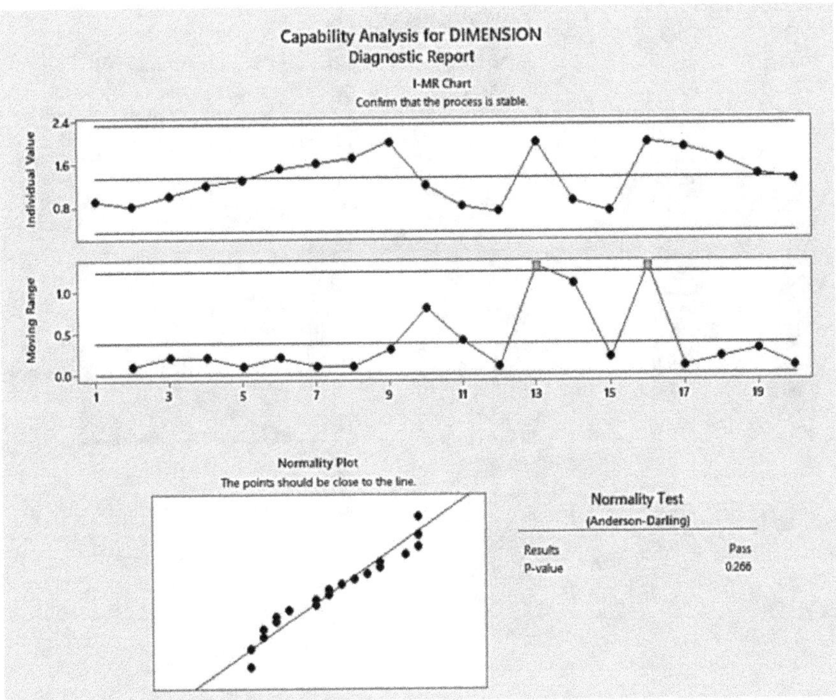

Figure 6.5 Process capability report.

as customer experience. Potential capability is obtained when there is no change or shifting in the process. As per the report generated by Minitab (Figure 6.5), the total range of out-of-specification value products was estimated at approx. 60.91%. From the process capability analysis curve, the process is stable to carry out. Z-bench value of process was –0.28 and the present DPMO level was 609099 which was very high in numbers simultaneously and has opportunities for improvement.

6.3.4 Improve

In the improvement phase, the main focus was given to resolve the main issues that arise during the process handling phase based on the results obtained from the capability index. In this improvement phase, several countermeasures have been taken in favor of improvement; some of these are mentioned below.

- Check points were included in the operation procedure standard.
- An associate induction programme was provided.
- The diameter of the holding pin was maintained.

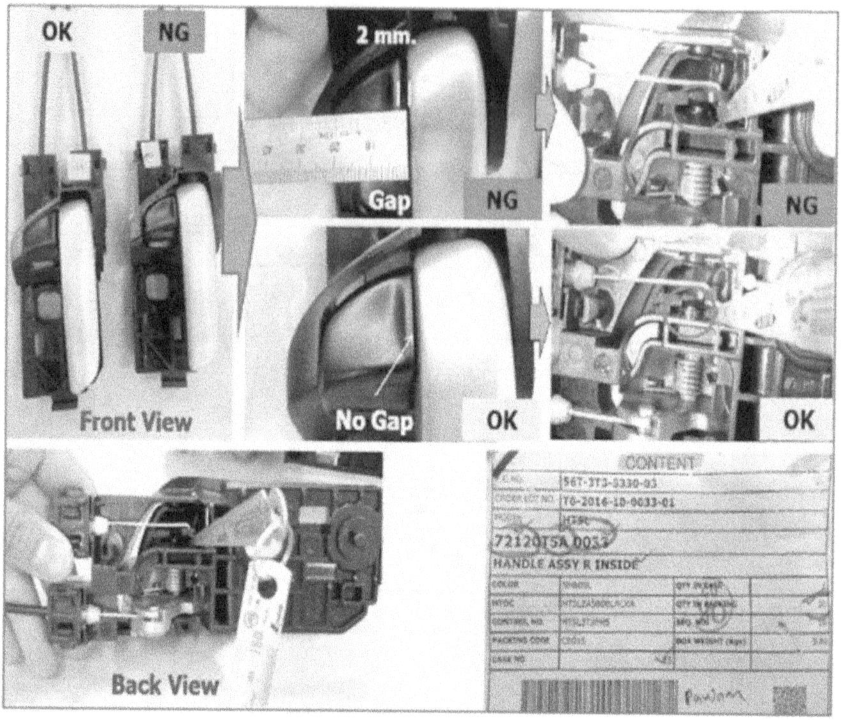

Figure 6.6 Door lever specifications.

A countermeasure had been taken, which might cause a decrease in the rejection rate (Figure 6.6).

6.3.5 Control

In the control phase, P-control charts have been plotted to specify the occurrence of assignable causes. MR control charts have been drawn, which show the stability of the process. The *p*-values are always used to validate the hypothetical test and measure the significance of the test. Figure 6.7 shows the P-control charts of the car door handle before and after the countermeasure process. I found it much more stable. For a hypothetical test, the process is more significant when the *p*-value is lower. In the present analysis, the *p*-value was decresed from 0.266 to 0.253, which also validates that the procedure followed during the process is more significant with a new Z bench value of 1.59 from –0.17 as shown in Figure 6.8.

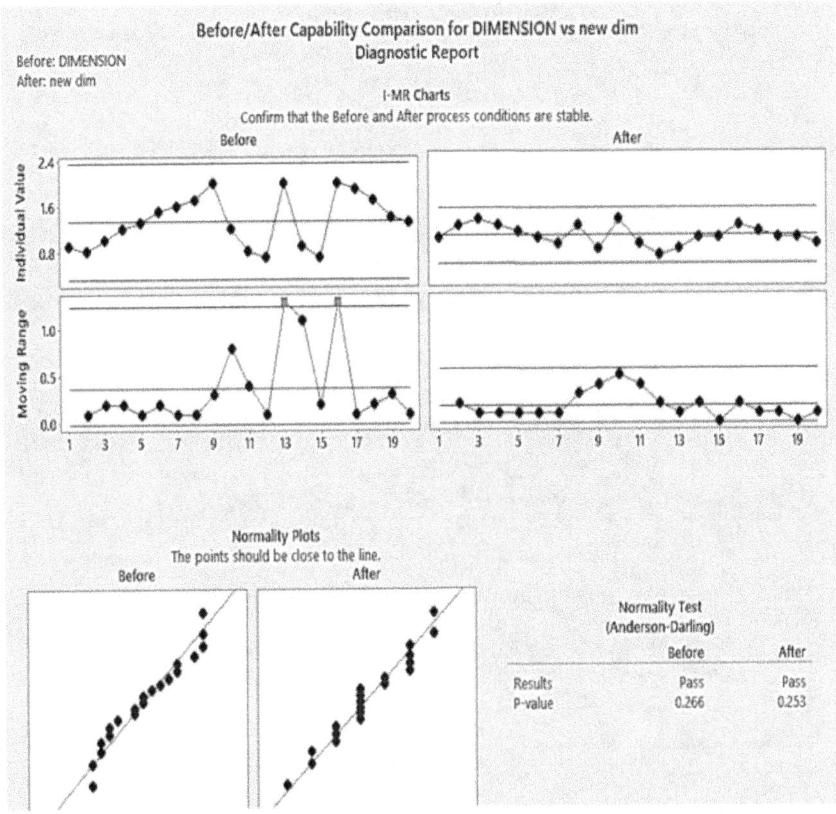

Figure 6.7 Control chart before/after counter measures.

6.4 CONCLUSIONS AND FUTURE SCOPE

This represents the findings and limitations of the present study and the scope for future research. Motivated by the observed gap between hypothetical and practise implementation of six-sigma deployment, this study explains the importance of empirical studies like expert surveys in developing countries like India. The myth that six-sigma is limited to a specified large industry or organization is abolished through this case study. Implementation of six-sigma in automobiles is very tough but not impossible. Several factors are required, and six-sigma techniques can be used. This case study will provide great knowledge to new researchers, scholars, and six-sigma beginners and contribute to society as well. The main findings of the present study are listed below:

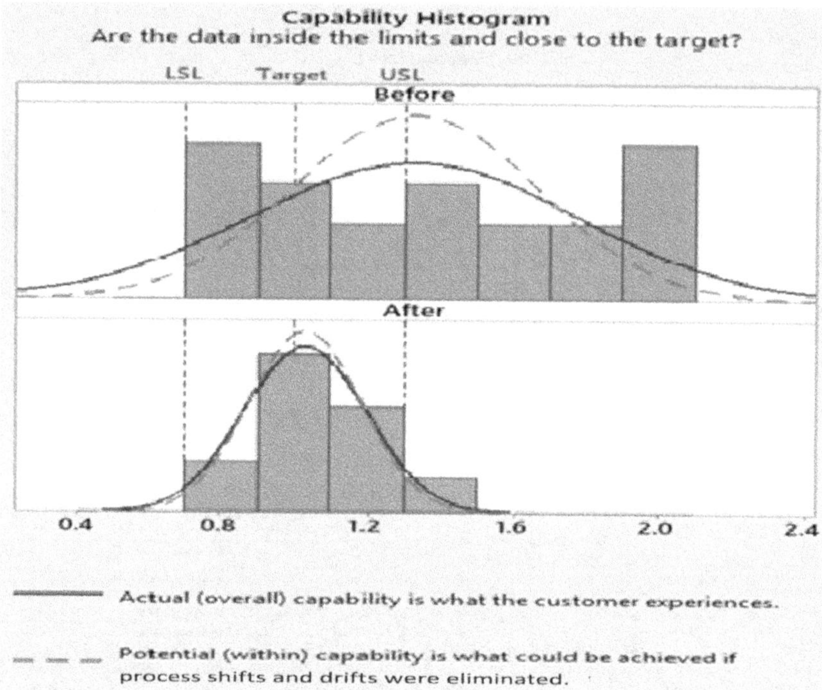

Process Characterization

	Before	After	Change
Total N	20	20	
Subgroup size	1	1	
Mean	1.33	1.03	-0.3
StDev(overall)	0.45434	0.16890	-0.28544
StDev(within)	0.33595	0.15398	-0.18197

Capability Statistics

	Before	After	Change
Actual (overall)			
Pp	0.22	0.59	0.37
Ppk	-0.02	0.53	0.55
Z.Bench	-0.28	1.40	1.68
% Out of spec (obs)	45.00	0.00	-45.00
% Out of spec (exp)	60.91	8.03	-52.88
PPM (DPMO) (obs)	450000	0	-450000
PPM (DPMO) (exp)	609099	80314	-528785
Potential (within)			
Cp	0.30	0.65	0.35
Cpk	-0.03	0.58	0.61
Z.Bench	-0.17	1.59	1.76
% Out of spec (exp)	56.60	5.58	-51.01
PPM (DPMO) (exp)	565954	55804	-510150

Figure 6.8 Capability histogram.

1. The gap between the knob and lever portions of the car door handle has been reduced using brainstorming and know-how techniques.
2. The *p*-values have been reduced from 0.266 to 0.253, which enhances the process significantly and also the level of product satisfaction.
3. The sigma level has also been increased by achieving the new Z bench value of 1.59 from –0.17 by reducing the DPMO values from 609099 to 80314.

REFERENCES

1. Singh, Mahipal, and Rajeev Rathi. "A structured review of lean six sigma in various industrial sectors." *International Journal of Lean Six Sigma* 10, no. 2 (May 29, 2019): 622–64. https://doi.org/10.1108/ijlss-03-2018-0018
2. Elboq, Raja, Mustapha Hlyal, and Jamila El Alami. "Empirical assessment of critical success factor of lean and six sigma in the Moroccan automotive industry." *IOP Conference Series* 827, no. 1 (April 1, 2020): 012043. https://doi.org/10.1088/1757-899x/827/1/012043
3. Bastos, Nuno Miguel Moreno Teixeira, Anabela Carvalho Alves, Felipe Castro, Júlio César Duarte, LuíS Pinto Ferreira, and F. J. G. Silva. "Reconfiguration of assembly lines using lean thinking in an electronics components' manufacturer for the automotive industry." *Procedia Manufacturing* 55 (January 1, 2021): 383–92. https://doi.org/10.1016/j.promfg.2021.10.053
4. Yang, Hui, Prahalada Rao, Timothy W. Simpson, Yan Lu, Paul Witherell, Abdalla R. Nassar, Edward W. Reutzel, and Soundar Kumara. "Six-sigma quality management of additive manufacturing." *Proceedings of the IEEE* 109, no. 4 (April 1, 2021): 347–76. https://doi.org/10.1109/jproc.2020.3034519
5. Gaikwad, Lokpriya, Umesh Bhushi, and S. N. Teli. "Implementation of six sigma methodologies to gain a competitive advantage: a case study approach." In *2022 Advances in Science and Engineering Technology International Conferences (ASET)* (February 21, 2022). https://doi.org/10.1109/aset53988.2022.9735103
6. Bewoor, Anand K., Mitali U. Kharul, Saloni G. Gosavi, and Tanvi J. Kuray. "Application of 7-step problem-solving methodology for defect elimination: a case study in an automotive industry." In *Lecture Notes in Mechanical Engineering* (2022): 697–707. https://doi.org/10.1007/978-981-16-9952-8_60
7. Kumar, V. Ravi, Vikas Modgil, and Ajay Kumar. "Stochastic petri nets modelling for performance assessment of a manufacturing unit." *Materials Today: Proceedings* 56 (January 1, 2022): 215–19. https://doi.org/10.1016/j.matpr.2022.01.073
8. Kumar, Ajay, Hari Singh, Parveen Kumar, and Bandar AlMangour. *Handbook of Smart Manufacturing*. CRC Press eBooks (2023). https://doi.org/10.1201/9781003333760

9. Ajay, Ajay, Ashwini Kumar, Parveen, Rajesh Kumar Goel, and Ravi Kant Mittal. *Waste Recovery and Management*. CRC Press eBooks (2023). https://doi.org/10.1201/9781003359784
10. Rani, Sangeeta, Khushboo Tripathi, and Ajay Kumar. "Machine learning aided malware detection for secure and smart manufacturing: a comprehensive analysis of the state of the art." *International Journal on Interactive Design and Manufacturing (IJIDeM)* (October 25, 2023). https://doi.org/10.1007/s12008-023-01578-0
11. Rani, Sangeeta, Khushboo Tripathi, Yojna Arora, and Ajay Kumar. "Analysis of anomaly detection of malware using KNN." In *2022 2nd International Conference on Innovative Practices in Technology and Management (ICIPTM)* (February 23, 2022). https://doi.org/10.1109/iciptm54933.2022.9754044
12. Joshi, Vishwas Deep, Priya Agarwal, and Ajay Kumar. "Fuzzy transportation planning: a goal programming tactic for navigating uncertainty and multi-objective decision making." *International Journal on Interactive Design and Manufacturing (IJIDeM)* (November 24, 2023). https://doi.org/10.1007/s12008-023-01634-9
13. Sathish kumar Arumugam, Sachin Kumar, Pramod Sridhara, Srinivasan Raju, Ashwin Prabhu Gnanasekaran, Nanthakumar Sivasamy and Thangarajan Sivasankaran Senthilkumar "Machine learning-based investigation of wear and frictional behavior in graphite-reinforced aluminum nanocomposites." *NanoWorld Journal* 9 (October 11, 2023). https://doi.org/10.17756/nwj.2023-s3-054

Chapter 7

Navigating the roadmap of intelligent transport in Indian smart cities

Sectorial analysis, challenges, and key elements of data science enabled solutions

Madhuri Patel, Samir B. Patel, Debabrata Swain, and Rajeev Gupta

Department of Computer Engineering, Pandit Deendayal Energy University, Gandhingar, India

7.1 REQUIREMENT OF DATA SCIENCE IN CITY PUBLIC TRANSPORT SYSTEM AND ITS ADVANTAGES

In the era of modern civic life, technology is essential for planning, implementation, and observation of various services provided to public. The significance of intelligent transport system data for strategic planning has been successfully reported [1–3]. In urban areas, a major issue prevails in that the public still use private cars to travel, which increases congestion, reduces parking area, and increases environmental problems. Eventually this enhances the transport system under continuous strain [4]. Post-2000 AD, the world focuses on greenhouse gas emissions, and energy consumption in the transport sector. To be able to prevent this, completely shifting towards alternative green fuels is essential [5]. To overcome such issues, a robust and organized city public transport system, particularly bus transport, can reduce congestion and pollution and also reduces usage of personal vehicles [6]. By providing on-demand services in a timely manner with organized routes can increase customer satisfaction and eventually results in enormous public transport usage. There is much scope for advanced research particularly in growing countries. Such an improvement can be done by short-term forecasts of future demands with time series [7–11]. With the objective of minimum waiting times for passengers and minimum operating cost for authorities, it is essential to have an accurate forecasting of future demands with all affecting features [12–13]. All types of associated predictions can only be achieved by collecting all possible data in continuous form with the least human intervention.

The public transportation system in Indian cities confronts a number of technological and implementation issues that impede its efficiency,

DOI: 10.1201/9781003502470-7

effectiveness, and overall service quality. Many cities lack the infrastructure required for an efficient public transportation system. Inadequate fleet size, bus stops, a lack of dedicated bus lanes, ICT-enabled infrastructure, and limited park-and-ride facilities are all examples. Many cities' public transportation fleets are comprised of old and poorly maintained buses and other vehicles. Maximum bus utilization, limited maintenance frequency due to a lack of buses, eventually results in increased maintenance costs, more breakdowns, and a lower quality of service. To meet the demands of large and heavily populated urban regions, public transportation routes and services are sometimes insufficient. Many residential areas continue to be underserved or inaccessible to public transportation. In many cities, inconsistent and unreliable schedules are a prevalent problem. Passengers are frequently required to wait for extended periods of time without knowing when the next bus or train would come. In recent years, certain localities have deployed intelligent transportation systems. However, it was eventually limited to merely visual dashboard information that collects data and provides it on the dashboard. Further scheduling, analysis, and a shared platform are still not available in a satisfactory manner. Platforms for mandated data exchange are also less functional. The study becomes more complicated as more agencies are involved and their contracts are not uniform. We also suggest a globally accepted data structure for analysis, as well as its integration with other aspects, so that the system can eventually be steered more efficiently through the use of data science.

The Smart Cities Mission aimed to create sustainable infrastructure nationwide, focusing on urban infrastructure renewal and retrofitting ingenuity. Its goal was to improve citizens' quality of life and enhance civic participation through the implementation of smart solutions. This involved the development of resilient and sustainable urban infrastructure, including affordable housing, green energy, clean water, and efficient waste management. Given the growing migration to large cities, it became crucial to establish efficient and eco-friendly public transportation systems and robust digitalization for safe commuting and better governance. A total of 100 cities were identified as smart cities in different phases, considering various parameters. Figure 7.1(left) shows the highest fund allocation in area development, urban transport mobility, and energy sectors, as they cater to the daily needs of a large population [14].

Public transport in large cities consists of local bus transport, dedicated bus rapid transport, metro, or local trains. Currently, we focus on bus transport system where the Figure 7.1(right) shows the total number of cities with local public transport facilities. The further bifurcation is provided with local transport, BRTS transport consists of intelligent transport system (ITS) system. Maharashtra, Karnataka, Gujarat, and Madhya Pradesh have cities consisting of the highest above facilities. However, Delhi, Mumbai, Bengaluru, Ahmedabad, and Thane are largest public commute using public transport

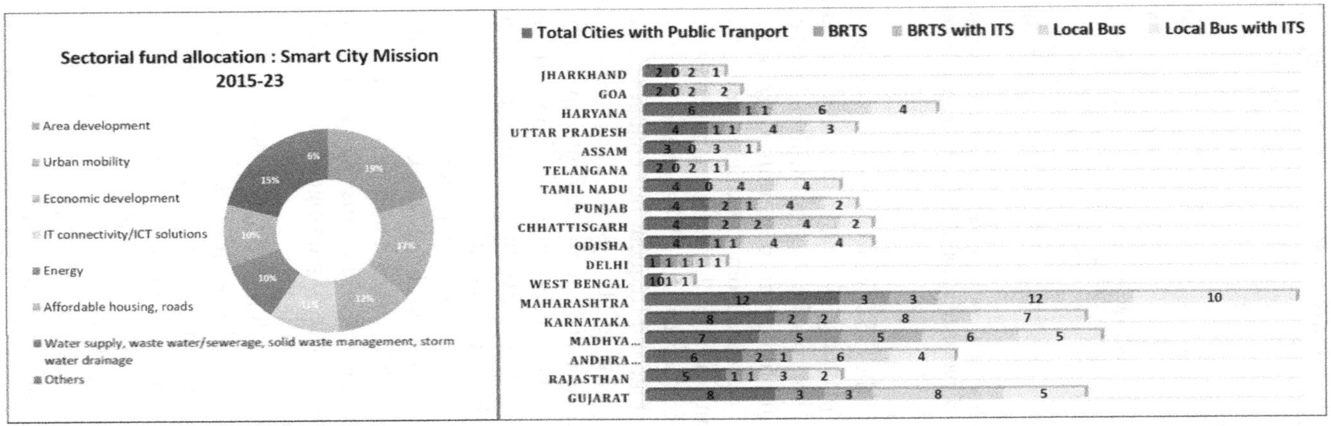

Figure 7.1 (Left) Fund allocation of various sectors of smart city mission. (Right) The state-wise city public transport system distribution with dedicated corridor and ITS-enabled system.

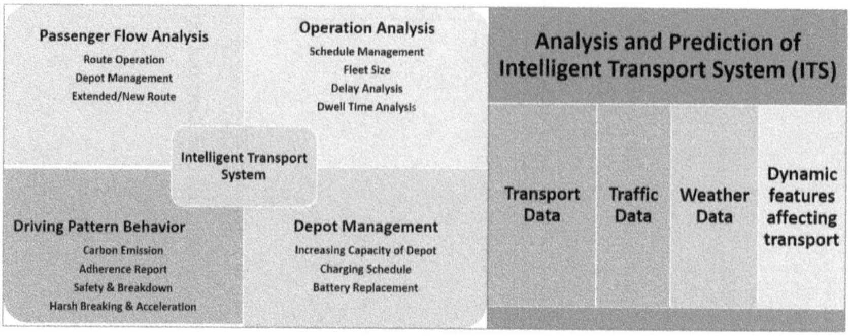

Figure 7.2 (Left) ITS features and various parameters of study. (Right) Integration of various pieces of information for effective outcomes of the ITS system.

system. The objective of the ITS system is to provide stakeholders with accurate information for route planning, passenger flow management, and optimizing fleet size. This leads to improved operations with reduced emissions.

Reliability is vital in public transportation. Currently, bus location data from GPS, passenger flow data from AFC, ticket or e-ticket details, and staff and route allocation information from the management portal provide data with varying levels of accuracy [15]. Negative impacts such as bus delays, inefficient route planning, and frequent bus changes can be observed.

To effectively implement an ITS system, data science plays a crucial role in areas such as passenger flow analysis, route operation analysis, driving pattern behavior, and depot management, as shown in Figure 7.2(left). This article focuses on various reviews that address these issues and essential features required for the implementation.

In day-to-day operations, various parameters directly impact transportation behavior. These include weekdays/weekends, holidays (religious/non-religious), time slots (peak hours and non-peak hours), weather conditions (rainfall intensity), and unexpected events (pandemic, cyclone, earthquake, elections, etc.). The magnitude of passenger flow dependency varies across different routes and areas of the city.

The implementation of machine learning technologies in the field of public transportation data analysis has enormous promise for India. With its rapidly expanding urban population and increasing transportation issues, there is an urgent need to reorganize existing networks. Machine learning, with its ability to process massive information and draw relevant insights, has the potential to significantly enhance this area. Being a major attractive area of the research, there are ample works already reported in numerous area of ITS. Despite the same, by seeing dynamism of the field the minor improvement in accuracy will result in a more reliable system.

The major issues of the public transport system in India are as follows:

- Planning of routes and covering of locations
- Management of routes with passenger flow
- Avoidance of bunching of buses at station
- Information dissemination for Public Private Partner
- Delay in routes
- Human resources planning
- Enhancement of operational efficiency
- Optimizing operating costs
- Establishing intelligent information management systems
- Quick emergency response

7.2. OVERVIEW OF DATA SCIENCE AND IMPLEMENTATION ON VARIOUS FEATURES OF SMART CITIES

Currently, the research work on various parameters of ITS as mentioned in Figure 7.2(left) have been carried out with various algorithms with single or hybrid model consists of traditional algorithms, regression models, machine learning models, and deep learning models. Researchers have reported various fields of study based on various features, such as its location, implantation methodology used for prediction of the passenger flow, or trip time estimation. The supporting data as mentioned in Figure 7.2(right) as per the features affecting research has been integrated from various authorities.

Intelligent transport system is wide area of research where the information dissimilation dashboard, passenger flow and their travel behavior, bus allocation, trip time, dwell time and automatic scheduling is important. Recently, with dealing of big data and GIS information, study related to development of visual analytics, dashboard has been reported [16–18]. During the initial development of such features, the Kalman filter is the most viable tool for analysis. The study has been reported for passenger flow prediction [19] and trip time estimation [20, 21]. Passenger flow prediction provides vital information for planning of services. The travel behavior and area-wise commutation have been reported based on travel card, AFC, and travel history [22–25]. This is vital information for additional routes and scheduling. Machine learning and time-based neural network algorithm provides exceptional accuracy for a dedicated route of transport like BRTS, urban rail, etc. [26–31]. The estimation of passenger demand also related various parameters like weather condition, day of commute, holidays, origin, and destination. Prediction of flow despite such influenced external parameters can be handled with data science [32–35]. Integrated work including passenger flow and bus travel time with various methods have also been studied [36, 37]. Bus travel time with various implementation models provides accurate information like arrival time, trip time, dwell time, next scheduling time based on GPS data, and other related information [38–51].

Day-to-day requirement of the civic society changes from the purpose of traveling, time of traveling, area of living etc. The existing reported work supports the development of new features in public ITS system for city transport. But at the same time it creates new research opportunities. We have identified issues related to data collection and methodology, listed lacuna in existing implementation and also proposed how ITS can be extended to new pathways for public transport planning.

7.3 DISCUSSION ON THE IMPLEMENTATION STAGES OF ANALYTICS FOR SMART CITIES

In India, limited work has been reported on advanced implementation of smart cities using data science, particularly in the field of ITS. This could be attributed to the scarcity of available data and the lack of integration among various authorities' databases. Currently under smart city project various databases have been recorded like transport data, traffic light violation, air pollution measurement, traffic flow, smart signal, vehicle registration, etc.

Figure 7.3 outlines different divisions for an effective ITS system in smart city projects, which can be studied and focused on to provide comprehensive transportation solutions. It is important to note that this pertains to one of the largest government spending initiatives, directly impacting around

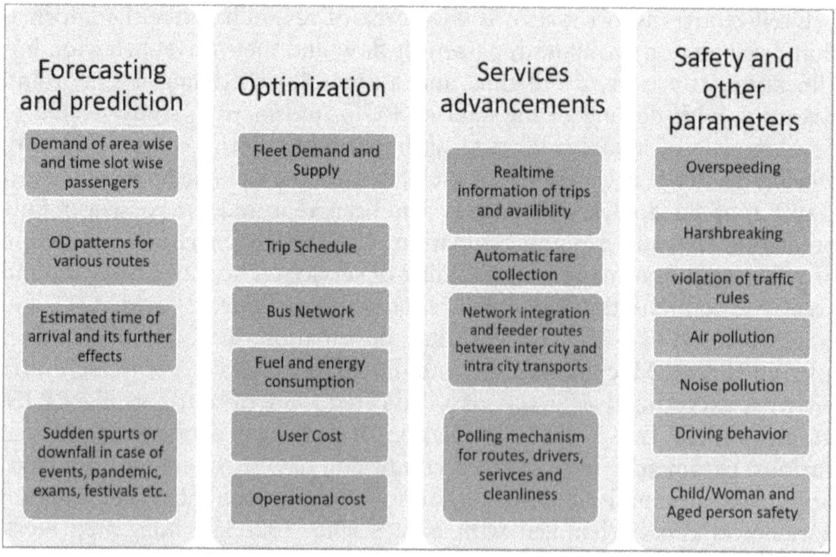

Figure 7.3 Features and division of studies and implementation for city public transport.

2.5 billion people. Machine learning, deep learning, and visual analytics methods can effectively analyze these features, which may include numerical, sensor-based, image-based, and other raw data formats. Typically, projects under this mission are implemented through special purpose vehicles (SPVs) coordinated by knowledge partners who work with multiple agencies. The initial step involves establishing a common database support system with minimal suggested features and creating a common open data source platform. Currently, data sharing platforms are operational, but the goal is to eventually transition them into open-access and reliable resources for researchers.

7.4 SUGGESTIVE METHODS AND FEATURES FOR INTEGRATED COMMAND AND CONTROL CENTERS (ICCC)

Initially, basic linear models were used for transport system planning. However, rapid urban growth, factors like traffic congestion, weather conditions, multiple routes, interconnected systems (e.g., Metro trains, dedicated corridors, Bus Rapid Transit), and changing travel purposes have made the dynamics more complex. Consequently, the system now incorporates dynamic transport information, estimated travel times, intelligent scheduling, automatic fare cards, GPS tracking, passenger behavior analysis, and decision-making for bus travel and transitioning to electric vehicles. To support these developments, improved databases and information on bus travel details and passenger flow behavior are essential. Figure 7.2(right) illustrates various sources of information required for analytics and prediction, which influence passenger flow patterns and bus travel behavior. Table 7.1 suggests a minimal dataset integrating various transport fields, which can be utilized for visual analytics, prediction, and optimization. The process is interdependent and integrated, where accurate and timely information is crucial for reliable outputs. This approach helps reduce passenger waiting times, enhance resource utilization, and improve dynamic scheduling.

Bus trip time prediction, dwell time, and passenger flow have received significant attention in various predictive studies.

Initially, statistical models such as auto regressive integrated moving average (ARIMA), seasonal auto regressive integrated moving average (SARIMA), and Kalman filtering were commonly used for linear models. However, considering the impact of various parameters on travel time and passenger flow, it is likely that real data exhibits non-linearity or significant variation.

In such cases, non-linear methods such as support vector regression (SVR), artificial neural networks (ANNs), back propagation neural networks (BPNNs), extreme learning machines (ELMs), and deep belief networks

Table 7.1 Various fields and available data corresponding to the division with purpose

Field of data	*Types of data*	*Remarks*
Transport data (system and human resource information)	Depot name and number	Reflects various depot of city
	Driver name and number	Reflects driver assigned to depot
	Conductor name and number	Reflects conductor assigned to depot
	E-ticket devices	Device number allocated to conductor for particular route
	AFC device	Device number allocated to waypoint for issuing tickets
Transport data (bus and trip details)	Vehicle number	Bus number
	Route number	Unique OD route number
	Trip direction	To and Fro indication of trip
	Covered waypoints	Total number and list of waypoints of a route
	Trip timing	Trip starting and end time of route
	Trip distance	Distance traveled during trip
	Speed	Speed of bus during trip with timestamp
	Harsh breaking	Acceleration and deceleration details with timestamp
	Location of bus	Longitude and latitude of bus with timestamp
Transport data (ticket details)	E-ticket issued for a route or OD	Reflecting number of passengers boarding for particular destination
	Origin and Destination of trip	Reflecting traveling of passenger
	Ticket issue date and time	Reflecting ticket issue time
Traffic details	Image of boarding station	Density of passenger at particular time for head count (demand)
	Real-time traffic position	Congestion at particular location with time
Weather and pollution data	Day length	Time between sunrise and sunset
	Temperature	Temperature for seasonal effect
	Rain fall	Understanding flow and trip time with various rain condition from no rain to heavy rain
	Air pollution	Visibility and quality of traveling
	Noise pollution	Identifying noise level during travel
Dynamic information affecting transport	Day of week	Travel day of week indicates weekend/weekday
	Holiday	Indicates holiday of given date
	Events	Marking events which increase or decrease passenger flow like exam, social gathering, election, school vacation, cricket match, etc.
	Disaster (natural or artificial)	COVID-19, Taukte, earth quake, Biporjoy, accident zone, repair zone

(DBNs) were implemented. Random factors like weather, traffic congestion, holidays, and weekdays/weekends can affect bus trips and passenger behavior, as discussed earlier.

Artificial neural networks mimic intelligent information processing akin to the human brain, encompassing association, learning, memory, and pattern identification. However, neural network models have limitations such as computation complexity, over fitting, and a black box nature. Against the poor applicability and lower accuracy of multiple regression model, the neural network model achieves well in nonlinear systems provided the model is trained well with large historical dataset. The core principle of ensemble learning lies in integrating diverse sets of learners to enhance prediction accuracy. This technique focuses mainly on strengthening weak learners and combining multiple learners generated by the same algorithm. This leads to higher prediction accuracy. Furthermore, ensemble models have the ability to address challenges that an individual model may struggle with. One such example is dynamic and stochastic nature of transit running time in city public transportation. By leveraging the strengths of different models, ensemble learning provides an improved solution for complex and challenging prediction scenarios.

Based on the recent studies, Table 7.2 provides a summary of the focused areas and corresponding implementation methods. Additionally, advantages and disadvantages are listed in Table 7.3 to aid in proper analysis identification. We propose a minimal and standardized data structure for the ITS system, which can be easily implemented and migrated to all cities with minimal effort. These tables serve as a foundation for pre-processing and exploratory data analysis (EDA) to determine the significance and weightage of each feature in prediction. Table 7.4 presents a structure for passenger flow prediction, where certain parameters are encoded to remove value dependency.

By conducting EDA and predictions using this data, factors such as demand and supply, route planning, optimization, operational costs, feeder routes, station infrastructure, scheduling, and fleet management can be analyzed.

Along with passenger flow prediction and its usage for different fleet management, the travel time and dwell time prediction is also important for scheduling and improvising services pertaining to city public transport. Dataset features have also been identified and listed in Table 7.5 for implementation of travel and dwell time prediction and EDA. The time series models like LSTM, RNN, etc., are most appropriate. Currently, around 88 out of 100 smart cities have integrated command and control centers for data processing, analysis, and visualization. These features aid in planning, performance evaluation, and providing clear visuals of the entire intelligent

Table 7.2 Prediction and optimization field with various models

Prediction / Forecasting	Passenger analysis: trip and passenger demand, origin to destination travel pattern analysis	SVR, MLP [52], MLP, decision tree, naive Bayes, bagging, polynomial regression, random forest, SVM, XGBoost [53], linear regression, LSTM [54], SVM, KNN [55], random forest, XGBoost [56], linear regression, XGBoost [57], DCNN [58,59], lightGBM [60]
	Bus driving analysis: driving behavior/harsh breaking analysis/ variation in speed/ estimated travel time/ dwell time/travel time	GRU, Seq2Seq [61], LSTM, KNN [62], linear regression, SVM, XGBoost [63], LSTM [64–69], LSTM, GNN [70],LSTM [71],GNN [72],K-MEAN, SVM [73], naive Bayes, KNN [74],Fuzzy, KNN [75], linear regression, SVM, KNN, GRU, LSM [76], LSM, DCNN [77], KNN [78], GNN [79–81], SVR, GRU [82–83], fuzzy [84–86], linear regression [87], K-mean [88], reinforcement learning [89], genetic algorithm [90]
	Safety: accident and travel safety	RNN, DCNN [91], MLP [92, 93], LSTM [94, 95], random forest [96–99], lightGBM, XGBoost, AdaBoost, random forest [100], K-mean, XGBoost [101–102], DCNN [103], lightGBM [104]
	Bus arrival time	MLP [31,48,105]
	Weather condition	MLP [106], SVM [107]
	COVID-19 pandemic	STL-LSTM [108]
	Air quality	MLP, XGBoost [109], support vector regression (SVR) [110]
Optimization	Transport demand and supply	K-mean [111]
	Overall scheduling	XGBoost [48,112]
	Transport network	MLP [113]
	Fuel consumption and efficiency	MLP [114–115]
	User cost	Regression, random forest, XGBoost [116]
	Operational cost	Random forest [117], MLP [118], others [117–122]

transport system (ITS) operation. In terms of cost optimization, service reliability, and public-private partnership (PPP) mode operations, these details and user-friendly dashboards are valuable for all stakeholders.

They also facilitate assessing the impact of implemented measures on the ITS system. After reviewing the peer literature, their available features for data science, operations and infrastructure and comparing with current situation of operations of city public transport in India we suggest the following points based on our experimental observations also:

Table 7.3 Advantages and disadvantages of various models

Category	*Commonly used models*	*Advantages*	*Disadvantages*
History of the same period	Smoothing method	Easy to compute and performance is good with linear data	Excessive dependency on linear patterns of historical data
Time series	Kalman filtering, AR (auto regressive), ARIMA	Efficient results in time series and also its interpretation	More error observed in non-linear data
Machine learning	SVM SVR, K-nearest neighbour, linear regression, decision tree, random forest	Performs well with non-linear data	Under performance in large data
Deep learning	RNN, LSTM, GRU	Suitable for linear and nonlinear learning patterns with decent data fitting capability	Computational complexity, black box model
Ensemble model	AdaBoost, Bootstrapped, aggregation, stacked generalization, gradient boosting machines, gradient boosted regression trees	Appropriate model selection in ensemble techniques for different data set	Overfitting, low interpretability, and higher error with unbalanced data

- Trip estimation: bus scheduling concerns, such as bunching or delayed trips, significantly impact daily operations. Using historical data, traffic patterns, and special events, machine learning algorithms can build more accurate and flexible timetables. Leveraging GPS and other vehicle sensors can give passengers real-time schedule updates, improving route planning. Bus stations and depots can use RFID or ultrasound sensors to estimate arrival and departure times. This will clarify bus journey behavior. Machine learning can dynamically alter routes to traffic, accidents, and special events, reducing travel time. For more accurate results, integrate and supply this data unanimously. LSTM, RNN, and boosting models estimate journey duration well, according to the review. Boosting models to deploy more data and better learning can yield the greatest outcomes.
- Passenger flow management: machine learning algorithms predict demand for routes and periods, enabling optimal resource allocation by transport authorities. Dynamic pricing based on demand and occupancy encourages off-peak travel. The examined regions

Table 7.4 Various fields and available data corresponding to passenger flow prediction

Feature	*Type of feature*	*Value set*	*Remarks*
Route number	Discrete Numeric	One Hot Encoding	Encoded
Route direction	Discrete Numeric	{0,1} One Hot Encoding	Encoded
Boarding station number	Discrete Numeric	One Hot Encoding	Encoded
Day	Discrete Numeric	{0,1,2,3…5,6}	Mapped using pandas day of week
Month	Discrete Numeric	{1,2,3,4…11,12}	Mapped using pandas month
Year	Discrete Numeric	{2021,2022}	Mapped using pandas year
Time of boarding	Discrete Numeric	{08:15:23, 23:24:27,….}	Actual time of boarding
Holiday	Discrete Numeric	{0,1}	Mapped Holiday:0, Non-holiday:1
Event	Discrete Numeric	{0,1,2,3..}	Event may be rated with higher number higher expecting boarding than normal day
Rainfall	Discrete Numeric	{0,1,2}	Mapped No Rain:0, Moderate Rain:1, Extreme Rain:2
Day length	Continuous Numeric	{length of day}	No mapping required
Number of waypoints in particular route	Discrete Numeric	{1,2,3,4…}	No mapping required
Type of passenger	Discrete Numeric	{1,2,3,4,5..}	Mapped Normal:1, Student pass:2, office pass:3, old age:4, widow:5
Waypoint location identity	Discrete Numeric	{1,2,3,4,5..}	Mapped Residential:1, bus/railway:2, station:3,airport:4,commercial:5, entertainment place:6, educational area:7

Table 7.5 Various fields and available data corresponding to trip analysis and ETA and dwell time prediction

Feature	*Type of feature*	*Value set*	*Remarks*
Track I'd	Discrete numeric	One Hot Encoding	Primary Key
Vehicle number	String	One Hot Encoding	Unique Bus Number Plate
Vehicle I'd	Discrete numeric	One Hot Encoding	Unique vehicle number assigned for transport (foreign key)
Group I'D	Discrete numeric	One Hot Encoding	Designated depot number assigned to bus
Group name	Text	One Hot Encoding	Designated depot name assigned to bus
Track date and time	Date and time	{08:15:23, 23:24:27,....}	Date and time at which live location of vehicle measured
No. of waypoints in route	Discrete numeric	{1,2,3,4...}	No mapping required
Longitude	Discrete numeric	One Hot Encoding	Longitude co-ordinate of current station
Latitude	Discrete numeric	One Hot Encoding	Latitude co-ordinate of current station
Waypoint location identity	Discrete numeric	{1,2,3,4,5..}	1 for residential, 2 for bus/railway station, 3 for airport, 4 for commercial premises, 5 for picnic or entertainment place, 6 for educational area etc.
Speed	Discrete numeric	{10,12,...} actual number	Define the exact speed of bus
Ignition	Integer[0,1]	{0,1}	Define OFF "0" and ON "1"
Route flag	Integer[0,1]	{0,1}	Define To "0" and Fro "1"

will benefit more from OD passenger flow analysis for planning and implementation. Recent study using random forest, lightGBM, and neural network will improve predicting. Indian holidays are based on local calendar, not Gregorian calendar, hence event-based prediction requires additional data for training. Accurate categorical prediction models yield more reliable outcomes. Data from traffic sensors, passenger counts, and meteorological conditions can inform decision-making.

- Machine learning can predict vehicle and infrastructure maintenance needs, decreasing downtime and improving reliability.
- Enable seamless transitions between public transport modes with integrated ticketing and route planning.
- Encourage collaboration and data exchange among transit operators to enhance citywide services.

7.5 OUTCOME OF DATA SCIENCE ENABLED SOLUTION ON ITS

Machine learning and data analytics can be immensely beneficial in resolving challenges with Indian city public transportation systems. India has some of the most densely inhabited cities in the world, which frequently endure congestion, ineffective transit networks, and pollution. Authorities can make informed decisions, improve efficiency, improve user satisfaction, reduce congestion, and encourage sustainable and accessible transportation options by using machine learning and data science into the Indian city public transit system. Furthermore, data-driven solutions can assist in lowering operational costs and optimizing resource allocation. This research vision outlines a detailed plan for addressing the issues presented by Indian city public transportation systems by leveraging machine learning and data analytics. It describes the research's objective, goals, methodology, and projected conclusions, as well as its relevance and ethical considerations. This review article also offers a systematic framework for utilizing data science tools in intelligent transport systems (ITS) in India. It provides insights into local transport breakdown, the on-going implementation of ITS in specific cities, and the integration of databases in smart city projects. The study emphasizes the significance of reliable transport services and suggests steps for pre-processing, exploratory data analysis (EDA), and evaluating prediction models for trip time analysis and passenger flow. These findings contribute to addressing existing challenges and proposing measures to meet the evolving needs of ITS in smart cities, ultimately ensuring reliable transport services.

REFERENCES

[1] Pelletier M.-P. et al., "Smart card data use in public transit: A literature review", *Transportation Research Part C: Emerging Technologies* (2011), 19(4), 557–568.

[2] Moreira-Matias L. et al., "Improving mass transit operations by using AVL-based systems: A survey", *IEEE Transactions on Intelligent Transportation Systems* (2015), 16(4), 1636–1653.

[3] Furth, P. G. et al., "Using archived AVL-APC data to improve transit performance and management", *Washington: Transit Cooperative Res program (TCRP) Report 113, Transp Res Board* (2006), 45–58.

[4] UITP, "Demographic changes and challenges for regional transport", *International Association of Public Transport* (2013), 105–235.

[5] Chen G. and Kauppila, J., "Global urban passenger travel demand and CO2 emissions to 2050: New model", *Transportation Research Record* (2017), 2671.1, 71–79.

[6] Li D. et al., "A spatio-temporal structured LSTM model for short-term prediction of origin-destination matrix in rail transit with multisource data", *IEEE Access* (2020), 8, 84000–84019.

[7] Liang X. et al., "A deep spatio-temporal fuzzy neural network for passenger demand prediction", *In Proceedings of the 2019 SIAM International Conference on Data Mining*. SIAM (2019), 100–108.

[8] Liyanage, S. et al., An agent-based simulation approach for evaluating the performance of on-demand bus services. *Sustainability* (2020), 12, 4117.

[9] Liyanage, S. et al, "Flexible mobility on-demand: Evironmental scan", *Sustainable Times* (2019), 11, 1–39.

[10] Smith B. L. et al., "Comparison of parametric and nonparametric models for traffic flow forecasting", *Transportation Research Part C: Emerging Technologies* (2002), 10, 303–321.

[11] Zhou C., Dai, P., & Li, R., "The passenger demand prediction model on bus networks", *2013 IEEE 13th International Conference on Data Mining Workshops*, IEEE (2013, December), 1069–1076.

[12] Ma Z. et al., "Predicting short-term bus passenger demand using a pattern hybrid approach", *Transportation Research Part C: Emerging Technologies* (2014), 39, 148–163.

[13] Tirachini A. et al., "Crowding in public transport systems: Effects on users, operation and implications for the estimation of demand" *Transport Respiratory Part A Policy Pract* (2013), 53, 36–52.

[14] Annual Reports of MoHUA between FY 2018–19 and FY 2021–22. Available online at: https://mohua.gov.in/cms/annual-reports.php. Last accessed on 25 January 2023.

[15] M. Patel et al., "Visual analysis of thane intelligent transport system", *2022 International Interdisciplinary Humanitarian Conference for Sustainability (IIHC), Bengaluru* (2022), 58–64.

[16] Godfrid, J. et al., "Analyzing public transport in the city of Buenos Aires with MobilityDB", *Public Transport* (2022), 14, 287–321.

[17] Ang, Kenneth Li-Minn et al., "Emerging technologies for smart cities transportation: Geo-information, data analytics and machine learning

approaches", *ISPRS International Journal of Geo-Information* (2022), 11(2), 85.

[18] Sobral T et al., "Visualization of urban mobility data from intelligent transportation systems", *Sensors (Basel)* (2019), 19(2), 332.

[19] GONG M. et al., "Sequential framework for short-term passenger flow prediction at bus stop", *Transportation Research Record: Journal of the Transportation Research Board* (2014), 2417, 58–66.

[20] Kumar, B. A., Vanajakshi, L., and Subramanian, S. C., "Pattern-based bus travel time prediction under heterogeneous traffic conditions", Transportation Research Record, Transportation Research Board, National Research Council, Washington, DC (2013).

[21] Cong Bai et al., "Dynamic bus travel time prediction models on road with multiple bus routes", *Computational Intelligence and Neuroscience* (2015), 2015, 9.

[22] Tong Zhang et al., "Measuring positive public transit accessibility using big transit data", *Geo-spatial Information Science* (2021), 24(4), 722–741

[23] Ghasemlou K et al., "Exploring equity in public transportation planning using smart card data", *Sensors (Basel)* (April 26, 2021), 21(9), 3039.

[24] Lv Wanjun et al., "A bus passenger flow prediction model fused with point-of-interest data based on extreme gradient boosting" *Applied Sciences* (2022), 12(3), 940.

[25] Jiao, F. et al., "An improved STL-LSTM model for daily bus passenger flow prediction during the COVID-19 pandemic", *Sensors* (2021), 21, 5950.

[26] Liang Zou et al., "Passenger flow prediction using smart card data from connected bus system based on interpretable XGBoost", *Wireless Communications and Mobile Computing* (2022), 2022, 13.

[27] Liang X., Wang, G., Min, M. R., Qi, Y., and Han, Z., "A deep spatio-temporal fuzzy neural network for passenger demand prediction", *In Proceedings of the 2019 SIAM International Conference on Data Mining.* Society for Industrial and Applied Mathematics (2019), 100–108.

[28] Zhang Jinlei & Chen et al., "Cluster-based LSTM network for short-term passenger flow forecasting in urban rail transit", *IEEE Access* (2019), 7. 147653–147671.

[29] Yuxing Sun et al., "A novel wavelet-SVM short-time passenger flow prediction in Beijing subway system", *Neurocomputing* (2015), 166, 109–121.

[30] Gummadi Reshma & Reddy et al., "Analysis of passenger flow prediction of transit buses along a route based on time series", *Information and Decision Sciences. Advances in Intelligent Systems and Computing*, vol 701. Springer, Singapore. https://doi.org/10.1007/978-981-10-7563-6_4.

[31] Nagaraj N et al., "Passenger flow prediction in bus transportation system using deep learning", *Multimed Tools Applications* (2022), 81(9), 12519–12542.

[32] Ouyang Qi et al., "An LSTM-based method considering history and real-time data for passenger flow prediction" *Applied Sciences* (2020), 10(11), 3788.

[33] Lingxiang Zhu et al., "XGBoost-based travel time prediction between bus stations and analysis of influencing factors", *Wireless Communications and Mobile Computing* (2022), 2022, 25.

[34] Zhang et al., "Forecasting public transit use by crowdsensing and semantic trajectory mining: Case studies" *ISPRS International Journal of Geo-Information* (2016), 5(10), 180.

[35] Toqué, F., Khouadjia, M., Come, E., Trepanier, M., and Oukhellou, L., "Short & long term forecasting of multimodal transport passenger flows with machine learning methods", *2017 IEEE 20th International Conference on Intelligent Transportation Systems (ITSC)*, IEEE (2017), 560–566.

[36] Liu Y. et al., "A deep learning based architecture for metro passenger flow prediction", *Transport Research Part C Emerging Technologies* (2019), 101, 18–34.

[37] Han Yong et al., "Short-term prediction of bus passenger flow based on a hybrid optimized LSTM network" *ISPRS International Journal of Geo-Information* (2019), 8(9), 366.

[38] Paul Patrick F. Balbin et al., "Predictive analytics on open big data for supporting smart transportation services", *Procedia Computer Science* (2020), 176, 3009–3018, ISSN 1877-0509.

[39] M. U. Farooq et al., "GPS based public transport arrival time prediction", *2017 International Conference on Frontiers of Information Technology (FIT)*, Islamabad, Pakistan (2017), 76–81.

[40] Jayanth Raghothama et al., "Analytics on public transport delays with spatial big data", *In Proceedings of the 5th ACM SIGSPATIAL International Workshop on Analytics for Big Geospatial Data* (BigSpatial '16). Association for Computing Machinery, New York, NY, USA (2016), 28–33.

[41] Steven I-Jy Chien et al., "Dynamic bus arrival time prediction with artificial neural networks", *Journal of Transportation Engineering* (2020), 128 (5), 429.

[42] Bin Yu et al., "Bus arrival time prediction at bus stop with multiple routes", *Transportation Research Part C: Emerging Technologies* (2011), 19(6), 1157–1170, ISSN 0968-090X.

[43] Yang et al., "Bus arrival time prediction using support vector machine with genetic algorithm", *Neural Network World* (2016), 26, 205–217.

[44] Xu H. et al., "Bus arrival time prediction with real-time and historic data", *Cluster Computers* (2017), 20, 3099–3106.

[45] Marko Čelan et al., "Bus-arrival time prediction using bus network data model and time periods", *Future Generation Computer Systems* (2020), 110, 364–371, ISSN 0167-739X.

[46] Ma Jiaman & Chan et al., "Bus travel time prediction with real-time traffic information", *Transportation Research Part C: Emerging Technologies* (2019), 105, 536–549.

[47] Li Jinglin & Gao et al., "Bus arrival time prediction based on mixed model", *China Communications* (2017), 14, 38–47.

[48] Jiaman Ma et al., "Bus travel time prediction with real-time traffic information", *Transportation Research Part C: Emerging Technologies* (2019), 105, 536–549, ISSN 0968-090X.

[49] Niklas Christoffer Petersen et al., "Multi-output bus travel time prediction with convolutional LSTM neural network", *Expert Systems with Applications* (2019), 120, 426–435, ISSN 0957-4174.

[50] Hans, Etienne & Chiabaut et al., "Real-time bus route state forecasting using particle filter and mesoscopic modelling". *Transportation Research Part C: Emerging Technologies* (2015) 61, 121–140.
[51] Zhang, Kunpeng & Liu et al., "Short-term prediction of passenger demand in multi-zone level: Temporal convolutional neural network with multi-task learning", *IEEE Transactions on Intelligent Transportation Systems* (2019), 21.4, 1480–1490.
[52] Ma, T. et al., "Hybrid machine learning algorithm and statistical time series model for network-wide traffic forecast", *Transport Research Part C Emerging Technologies* (2020), 111, 352–372.
[53] Zhao X. et al., "Prediction and behavioral analysis of travel mode choice: A comparison of machine learning and logit models", *Travel Behaviour and Society* (2020), 20, 22–35.
[54] F. Ten et al, "A stepwise interpretable machine learning framework using Linear Regression (LR) and Long Short-Term Memory (LSTM): City-wide demand-side prediction of yellow taxi and for-hire vehicle". www.sciencedirect.com/science/article/abs/pii/S0968090X2030696.
[55] Jahangiri A. et al., "Applying machine learning techniques to transportation mode recognition using mobile phone sensor data", *IEEE Transactions on Intelligent Transportation Systems* (2015), 16, 2406–2417.
[56] Karnberger S. et al., "Network–wide prediction of public transportation ridership using spatio–temporal link–level informatio", *Journal of Transport Geography* (2020), 82, 102549.
[57] Wang S. et al., "Deep neural networks for choice analysis: A statistical learning theory perspective", *Transporation Research Part B Methodological* (2021), 148, 60–81.
[58] Sun P A. et al., "A stepwise interpretable machine learning framework using Linear Regression (LR) and Long Short-Term Memory (LSTM)". www.sciencedirect.com/science/article/abs/pii/S0968090X20306963?via%3Dihub
[59] Jahangiri A. et al., "Applying machine learning techniques to transportation mode recognition using mobile phone sensor data", *IEEE Transactions on Intelligent Transportation Systems* (2015), 16, 2406–2417.
[60] Karnberger S. et al., "Network–wide prediction of public transportation ridership using spatio–temporal link–level information", *Journal of Transport Geography* (2020), 82, 102549.
[61] Wang S. et al., "Deep neural networks for choice analysis: A statistical learning theory perspective", *Transporation Research Part B Methodological* (2021), 148, 60–81.
[62] Liu G. et al., "Passenger flow estimation based on convolutional neural network in public", *Transportation System Knowledge-Based System* (2017), 123, 102–115.
[63] Jin C.-J. et al., "The automatic detection of pedestrians under the high-density conditions by deep learning techniques", *Journal of Advanced Transportation* (2021), 2021, e1396326.
[64] Ha T.V. et al., "The application of gradient boost machine and local interpretable modelagnostic explanations in examining the travel multi-mode choice, the case study of Jakarta City, Indonesia", *Journal of Eastern Asia Society of Transportation Studies* (2019), 13, 503–522.

[65] Boukerche A. et al., "Performance modeling and analysis of a novel vehicular traffic flow prediction system using a hybrid machine learning-based model", *Ad Hoc Network* (2020), 106, 102224.
[66] Xianglong Luo, Danyang Li, Yu Yang, Shengrui Zhang, "Spatiotemporal traffic flow prediction with KNN and LSTM", *Journal of Advanced Transportation* (2019), 2019, 10. https://doi.org/10.1155/2019/4145353
[67] Chao Yu, Haiying Li, Xinyue Xu, Jun Liu, Jianrui Miao, Yitang Wang, Qi Sun, "Data-driven approach for passenger mobility pattern recognition using spatiotemporal embedding", *Journal of Advanced Transportation* (2021), 2021, 21. https://doi.org/10.1155/2021/5574093.
[68] Rusul L. Abduljabbar, Hussein Dia, Pei-Wei Tsai, "Unidirectional and bidirectional LSTM models for short-term traffic prediction", *Journal of Advanced Transportation* (2021), 2021, 16. https://doi.org/10.1155/2021/5589075
[69] Zhu H. et al., "A novel traffic flow forecasting method based on RNN-GCN and BRB", *Journal of Advanced Transportation* (2020), 2020, e7586154.
[70] Xiao, H. et al., "UB-LSTM: A trajectory prediction method combined with vehicle behavior recognition", *Journal of Advanced Transportation* (2020), 2020, e8859689.
[71] Bogaerts, T. et al., "A graph CNN-LSTM neural network for short and long-term traffic forecasting based on trajectory data", *Transport Research Part C Emerging Technologies* (2020), 112, 62–77.
[72] Xuan Zhao, Qiang Yu, Jian Ma, Yan Wu, Man Yu, Yiming Ye, "Development of a representative EV urban driving cycle based on a k-means and SVM hybrid clustering algorithm", *Journal of Advanced Transportation* (2018), 2018, 18. https://doi.org/10.1155/2018/1890753.
[73] Chen D. et al., "Driving style recognition under connected circumstance using a supervised hierarchical bayesian Model", *Journal of Advanced Transportation* (2021), 2021, e6687378.
[74] Boukerche, A. et al., "Artificial intelligence-based vehicular traffic flow prediction methods for supporting intelligent transportation systems", *Computer Network* (2020), 182, 107484.
[75] Bahuleyan H. et al., "Arterial path-level travel-time estimation using machine-learning techniques", *Journal of Computing in Civil EngIneering* (2017), 31, 04016070.
[76] Chuanxiang Ren, Chunxu Chai, Changchang Yin, Haowei Ji, Xuezhen Cheng, Ge Gao, Heng Zhang, "Short-term traffic flow prediction: A method of combined deep learnings", *Journal of Advanced Transportation* (2021), 2021, 15. https://doi.org/10.1155/2021/9928073
[77] Roy A. et al., "SST-GNN: Simplified spatio-temporal traffic forecasting model using graph neural network", *arXiv 2021*, arXiv:2104.00055.
[78] Ye, Jiexia, Furong Zheng, Juanjuan Zhao, Kejiang Ye, and Chengzhong Xu. "Incorporating reachability knowledge into a multi-spatial graph convolution based Seq2Seq model for traffic forecasting". arXiv preprint arXiv:2107.01528 (2021).
[79] Diehl F. et al., "Graph neural networks for modelling traffic participant interaction", *2019 IEEE Intelligent Vehicles Symposium (IV)* (219), Paris, France (9–12 June 2019), 695–701.

[80] Li Y et al., "A brief overview of machine learning methods for short-term traffic forecasting and future directions", *SIGSPATIAL Specification* (2018), 10, 3–9.

[81] Xiaolei Ru, Xiangdong Xu, Yang Zhou, Chao Yang, "Critical Segments Identification for link travel speed prediction in urban road network", *Journal of Advanced Transportation* (2020), 2020, 10. https://doi.org/10.1155/2020/8845804.

[82] Shengyou Wang, Chunfu Shao, Yajiao Zhai, Song Xue, Yan Zheng, "A multifeatures spatial-temporal-based neural network model for truck flow prediction", *Journal of Advanced Transportation* (2021), 2021, 14. https://doi.org/10.1155/2021/6624452

[83] Gao Y. et al., "Traffic speed forecast in adjacent region between highway and urban expressway: Based on MFD and GRU model", *Journal of Advanced Transportation* (2020), 2020, e8897325.

[84] Shang Q. et al., "An imputation method for missing traffic data based on FCM optimized by PSO-SVR", *Journal of Advanced Transportation, 2018* (2018), e2935248.

[85] Zhang, S. et al., "Speed Grade Evaluation of Public-Transportation Lines Based on an Improved T-S Fuzzy Neural Network"., *Journal of Advanced Transportation,* 2020 (2020), e8838084.

[86] Yongjie Xue, Rui Feng, Shaohua Cui, Bin Yu, "Traffic status evolution trend prediction based on congestion propagation effects under rainy weather", *Journal of Advanced Transportation* (2020), 2020, 14. https://doi.org/10.1155/2020/8850123

[87] Liu Z. et al., "A tailored machine learning approach for urban transport network flow estimation", *Transport Research Part C Emerging Technologies* (2019), 108, 130–150.

[88] Amin Mohammadnazar et. al, "Classifying travelers' driving style using basic safety messages generated by connected vehicles: Application of unsupervised machine learning", *Transportation Research Part C: Emerging Technologies* (2021), 122, 102917, ISSN 0968-090X, https://doi.org/10.1016/j.trc.2020.102917.

[89] Yanqiu Cheng, Chenxi Chen, Xianbiao Hu, Kuanmin Chen, Qing Tang, Yang Song, "Enhancing mixed traffic flow safety via connected and autonomous vehicle trajectory planning with a reinforcement learning approach", *Journal of Advanced Transportation* (2021), 2021, 11. https://doi.org/10.1155/2021/6117890.

[90] Yang S. et al., "Machine learning approach to decomposing arterial travel time using a hidden Markov model with genetic algorithm", *Journal of Computing Civil Engineering* (2018), 32, 04018005.

[91] Akhtar M. et al., "A review of traffic congestion prediction using artificial intelligence", *Journal of Advanced Transportation* (2021), 2021, e8878011.

[92] Chiou Y., "An artificial neural network-based expert system for the appraisal of two-car crash accidents", *Accident Analysis and Prevention* (2006), 38, 777–785.

[93] Siems-Anderson A.R. et al. "An adaptive big data weather system for surface transportation", *Transportation Research Interdiscipincary Perspective* (2019), 3, 100071.
[94] Lu Z. et al., "A new video-based crash detection method: Balancing speed and accuracy using a feature fusion deep learning framework, " *Journal of Advanced Transportation* (2020) , 2020, e8848874.
[95] Liu Z.-Q. et al., "Estimation of driver lane change intention based on the LSTM and Dempster–Shafer evidence theory", *Journal of Advanced Transportation* (2021), 2021, e8858902.
[96] Gan J. et al., "An alternative method for traffic accident severity prediction: Using deep forests algorithm", *Journal of Advanced Transportation* (2020), 2020, e1257627.
[97] Qiang Shang, Derong Tan, Song Gao, Linlin Feng, "A hybrid method for traffic incident duration prediction using BOA-optimized random forest combined with neighborhood components analysis, " *Journal of Advanced Transportation* (2019), 2019, 11. https://doi.org/10.1155/2019/4202735
[98] Jahangiri A. et al., "Applying machine learning techniques to transportation mode recognition using mobile phone sensor data", *IEEE Transactions on Intelligent Transportation Systems* (2015), 16, 2406–2417.
[99] Marks C. et al., "Identifying and labeling potentially risky driving: A multistage process using real-world driving data, " *Journal of Advanced Transportation* (2021), 2021, e8819094.
[100] Zhang X. et al., "An ensemble machine learning-based modeling framework for analysis of traffic crash frequency", *Computer-Aided Civil Infrastructure Engineering* (2020), 35, 258–276.
[101] Tang, J. et al., "Traffic incident clearance time prediction and influencing factor analysis using extreme gradient boosting model, " *Journal of Advanced Transportation* (2020), e6401082.
[102] Chen H. et al., "Analysis of factors affecting the severity of automated vehicle crashes using XGBoost model combining POI data", *Journal of Advanced Transportation* (2020), e8881545.
[103] Hesham M. Eraqi, Yehya Abouelnaga, Mohamed H. Saad, Mohamed N. Moustafa, "Driver distraction identification with an ensemble of convolutional neural networks", *Journal of Advanced Transportation* (2019), 2019, 12. https://doi.org/10.1155/2019/4125865
[104] Lai Y. et al., Bus arrival time prediction using wavelet neural network trained by improved particle swarm optimization. *Journal of Advanced Transportation* (2020), 2020, e7672847.
[105] Siems-Anderson A.R. et al., "An adaptive big data weather system for surface transportation", *Transportation Research Interdisciplinary Perspectives* (2019), 3, 100071.
[106] Jiandong Zhao, Hongqiang Wu, Liangliang Chen, "Road surface state recognition based on SVM optimization and image segmentation processing, " *Journal of Advanced Transportation* (2017), 2017, 21. https://doi.org/10.1155/2017/6458495
[107] Jiao F. et al., "An improved STL-LSTM model for daily bus passenger flow prediction during the COVID-19 Pandemic", *Sensors* (2021), 21, 5950.

[108] Wang A. et al., "Potential of machine learning for prediction of traffic related air pollution", *Transportation Research Part D: Transport and Environment* (2020), 88, 102599.

[109] Zhu C. et al., "Prediction on peak values of carbon dioxide emissions from the chinese transportation industry based on the SVR model and scenario analysis, " *Journal of Advanced Transportation* (2020), 2020, e8848149.

[110] Gocmen, E. et al., "Transportation problems for intermodal networks: mathematical models, exact and heuristic algorithms, and machine learning, " *Expert Systems Applications* (2019), 135, 374–387.

[111] Kang L. et al., "Urban traffic travel time short-term prediction model based on spatio-temporal feature extraction, " *Journal of Advanced Transportation* (2020), 2020, e3247847.

[112] Shmueli D., "Applications of neural networks in transportation planning", *Program Planning* (1998), 50, 141–204.

[113] Shams Shideh Shams Amiri et al., "Peeking inside the black-box: Explainable machine learning applied to household transportation energy consumption", *Computers, Environment and Urban Systems* (2021), 88, 101647, ISSN 0198-9715.

[114] Moradi E. et al., "Vehicular fuel consumption estimation using real-world measures through cascaded machine learning modeling, " *Transportation Research Part D Transport Environmental* (2020), 88, 102576.

[115] Budak A. et al., "Profit margin prediction in sustainable road freight transportation using machine learning, " *Journal of Cleaner Production* (2021), 314, 127990.

[116] Ben Beklisi Kwame Ayawli et al., "An overview of nature-inspired, conventional, and hybrid methods of autonomous vehicle path planning", *Journal of Advanced Transportation* (2018), 2018, 27.

[117] Kumar, Amit, Vinod Kumar, Vikas Modgil, and Ajay Kumar. "Stochastic Petri nets modelling for performance assessment of a manufacturing unit". *Materials Today: Proceedings* (2022), 56, 215–219.

[118] Kumar, Ajay, Hari Singh, Parveen Kumar, and Bandar AlMangour, eds. *Handbook of Smart Manufacturing: Forecasting the Future of Industry 4.0*. CRC Press (2023). https://doi.org/10.1201/9781003333760

[119] Kumar, Ashwini, Ravi Kant Mittal, and Rajesh Goel, eds. *Waste Recovery and Management: An Approach Toward Sustainable Development Goals*. CRC Press (2023). https://doi.org/10.1201/9781003359784

[120] Rani, Sangeeta, Khushboo Tripathi, and Ajay Kumar. "Machine learning aided malware detection for secure and smart manufacturing: A comprehensive analysis of the state of the art". *International Journal on Interactive Design and Manufacturing (IJIDeM)* (2023), 17, 1–28.

[121] Rani, Sangeeta, Khushboo Tripathi, Yojna Arora, and Ajay Kumar. "A machine learning approach to analyze cloud computing attacks". In *2022 5th International Conference on Contemporary Computing and Informatics (IC3I)*. IEEE (2022), 22–26.

[122] Joshi, Vishwas Deep, Priya Agarwal, and Ajay Kumar. "Fuzzy transportation planning: A goal programming tactic for navigating uncertainty and multi-objective decision making", *International Journal on Interactive Design and Manufacturing (IJIDeM)* (2023), 17, 1–29.

Chapter 8

Enhancing aircraft inspection through computer vision

Mitigating human error and improving safety

Vineet Bhatia and Sumati Sidharth
Defence Institute of Advanced Technology, Pune, India

8.1 INTRODUCTION

In the ever-evolving field of aviation, maintaining the highest standards of aircraft safety is of paramount importance. A crucial component of this safety protocol is regular and thorough inspection of aircraft exteriors. Currently, this inspection process relies heavily on human evaluators who meticulously assess the aircraft's external structure, looking for any signs of wear, damage, or foreign objects. Also, with increasing demands on aircraft utilization due to tight, maintenance operations face growing pressures to meet strict timelines, subsequently adding to workforce stress. However, due to the complexity and repetitive nature of these tasks, they are susceptible to human error, potentially compromising safety and operational efficiency [1]. Since the 1990s, research has been ongoing to incorporate robotic solutions for automating these external aircraft inspections [2, 3]. The primary objectives have been to support maintenance technicians in their diagnostic process and enhance the tracking of defects and damages in maintenance reports. Early robotic solutions revolved around inspecting the external surface skin with robots designed to crawl on the airplane [4]. Despite successful proofs of concept, certain limitations were identified that prevented full-scale deployment. At the start of the 2010s, a collaborative mobile robot named Air-Cobot was developed [5]. This wheeled robot can safely navigate around an aircraft, evading any obstacles in its path. The robot is equipped with two sensors specifically designed for inspection. The visual inspection is facilitated by a pan-tilt-zoom camera, which employs computer vision algorithms to analyze images captured at different points during movement or halts. Furthermore, the robot is equipped with a 3D scanner that can be elevated to examine the lower surface of the fuselage [6–8]. This advanced robot is capable of autonomous task execution or seamless collaboration with its operator. Alternative methods emphasize the use of cameras situated in the maintenance hangar for aircraft inspection [9]. In an increasingly reliant world on air travel, innovative strategies

DOI: 10.1201/9781003502470-8

are sought to improve aircraft inspections and reduce human error. Recent advancements in artificial intelligence (AI) and machine learning (ML), particularly computer vision, offer promising solutions to automate intricate tasks, enhance operational efficiency, and interpret visual information [10]. Computer vision-based inspection refers to the use of computer vision technologies to analyze and interpret visual information for the purpose of inspecting, assessing, or evaluating objects, systems, or processes. This approach involves the application of image processing, pattern recognition, and machine learning algorithms to interpret visual data captured by cameras or sensors [11].

8.2 AIRCRAFT VISUAL INSPECTION

The visual inspection of an aircraft's skin and substructure stands as a crucial stage in its maintenance process. This inspection aims to identify surface defects resulting from external factors such as lightning strikes, hailstorms, or regular wear from usage. Common defects detected include cracks, corrosion, dents, and damaged rivets. Currently, human technicians conduct visual inspections, which involve time-consuming practices, safety risks, and the possibility of human errors [12, 13]. Over 80% of inspections of a large transport category aircraft are visual inspections [14]. This involves meticulous examination of both interior and exterior aircraft surfaces to identify potential damages or irregularities that could negatively impact the aircraft's operation, structural integrity, or in-flight safety.

8.2.1 Categories of aircraft infection

Aircraft visual inspections are categorized into four distinct levels [15]

- **Level 1: walkaround.** This is a basic examination conducted from the ground level, aimed at identifying any obvious irregularities and assessing the overall condition and safety of the aircraft.
- **Level 2: general visual inspection** (GVI). This type of inspection involves a comprehensive check of the exterior, including selective hatches and openings, or the interior, when necessary, to discover any signs of damage, malfunctions, or anomalies.
- **Level 3: detailed visual inspection** (DET). This level of inspection entails a thorough visual check of a specific area, system, or assembly for signs of damage, breakdown, or non-uniformity. It may require the use of available inspection tools, surface treatment, and detailed access procedures.
- **Level 4: special detailed visual inspection.** This examination involves a thorough investigation of an identified component for installation or assembly to identify any damage, malfunctions, or anomalies. It

often requires the use of specialized method tools and equipment, and may involve complex disassembly and cleaning procedures.

8.3 AIRCRAFT VISUAL INSPECTION AND HUMAN ERROR

Several types of human errors can influence the effectiveness of aircraft visual inspection. These are briefly explained in subsequent paragraphs [16]:

- Perception errors [17]: these occur when inspectors misinterpret what they see. Such errors can be influenced by lighting conditions, visual illusions, fatigue, or even the individual's physical health.
- Memory errors: inspectors may forget the details of the objects they are supposed to inspect, especially in complex aircraft systems, leading to potential oversights.
- Decision errors: these involve making incorrect conclusions or judgments based on the information gathered during the inspection. For instance, an inspector might mistakenly identify a harmless anomaly as a serious defect, or vice versa.
- Skill-based errors: these can occur if the inspector lacks adequate training or experience, leading to an inability to conduct the inspection effectively or identify subtle anomalies.
- Attitudinal errors: inspectors' attitudes, such as complacency or overconfidence, can lead to overlooked or underestimated issues.

The utilization of AI techniques, particularly computer vision, is necessary to reduce human error in aircraft visual inspection. These capabilities contribute to enhanced aviation safety by minimizing the risks associated with human errors in inspection processes.

8.4 METHODOLOGICAL APPROACH

The research chapter uses a combination of literature review and practical use case studies to scrutinize the present status of computer vision for aircraft and engine inspection. The use cases have been incorporated from prominent aircraft manufacturers and solution providers. We conducted a comprehensive survey of the Scopus database in order to identify pertinent papers over the past decade, covering the period from 2013 to 2023. The keywords employed in the research context are 'computer vision', aircraft, engine', and 'inspection.' The Scopus search generated 199 documents related to aircraft and 132 documents related to engine. After a thorough examination of each of these documents, we successfully pinpointed 23 research papers directly relevant to aircraft and 15 research papers directly relevant to engine in the context of our subject of interest. Subsequently, we meticulously analyzed these research papers to gain valuable insights into the practical application of learning models for the detection of defects on air-craft surfaces.

Furthermore, as we sought a deeper understanding of the advancements in this field, we initiated an investigation into practical real-world applications, exemplified by use cases of these models, as demonstrated by prominent aircraft original equipment manufacturers (OEMs) and technology companies. This comprehensive investigation entailed a close scrutiny of the practical implementation of these models in defect detection and the automation of aircraft maintenance.

8.5 COMPUTER VISION FOR AIRCRAFT INSPECTION

The current methodology of aircraft and engine employs traditional manual inspection processes, where human inspectors visually assess and evaluate the aircraft and engine surface and components. This often involves subjective judgments and can be time-consuming, with the potential for human error. The process lacks automation and may not be scalable for large-scale inspections.

Computer vision technologies automate and enhance the inspection process. It involves the use of cameras or sensors to capture visual data, which is then processed using image analysis, pattern recognition, and machine learning algorithms. The proposed method aims to provide objective and consistent results, improving efficiency and accuracy in the inspection process. This technology allows for real-time feedback and can be customized for various inspection tasks, offering a more robust and scalable solution compared to the traditional manual approach.

The utilization of computer vision in the field of aircraft and engine inspection has seen significant advancements over the years. The journey can be traced back to as early as 1980, when experimental programs began using computer vision techniques to guide robot arms for aircraft assembly tasks (robotic drilling and riveting) [18]. These early systems demonstrated the ability to perform operations with reliability and accuracy in real time, marking a critical step in the integration of computer vision into the aerospace industry. The mid-1980s saw further exploration into the capabilities of computer vision, with researchers investigating the use of Fourier descriptors for the rapid recognition of two-dimensional shapes in computer vision systems [19]. These early investigations contributed to the development of algorithms that could be executed in a reasonable amount of time, even on the computing hardware available then. By 1988, the applications of computer vision systems had expanded to include the inspection of industrial parts, the navigation of autonomous vehicles, and the interpretation of aerial photographs [20]. The increased versatility of computer vision applications showcased the technology's potential to transform various aspects of the aerospace industry. As we moved into the 21st century, advancements in computer vision technologies continued to impact the aerospace industry significantly. Unmanned aerial vehicles (UAVs) equipped with advanced computer vision systems became vital tools for remote sensing observation. Advancements continued into the 2000s. In 2004, an automated system for drilling aircraft structures [21] was described, utilizing machine vision for hole position identification and feedback provision. In 2010, magneto-optical images [22] were used for aircraft rivet inspection. By 2011, a system was developed for quantitatively assessing aircraft surfaces [23], with a focus on painted surfaces of military aircraft.

In 2015, unmanned aerial vehicles (UAVs) integrated with high-resolution cameras were introduced for aircraft panel inspection [24]. The subsequent year (2016), a novel method was suggested for detecting and inspecting airplane tires. Tire detection is carried out using normalized correlation-based template matching, and the extraction of the tires from the background is accomplished through the GrabCut segmentation method [25]. By 2018, machine vision was being applied to inspect and evaluate aircraft coatings [26], critical for protecting aircraft skin from corrosion and erosion. In 2019, unmanned aerial vehicle (UAV) was utilized for inspection of aircraft exterior screw image. A technique employing a convolutional neural network (CNN) was applied to characterize zones of interest (ZOI) and extract screws from images [27]. In 2020 inspection of tapered interference-fit fasteners, such as Taper-lok, using computer vision was undertaken. [28]. Also, a novel two-step process was devised to automate the visual inspection of an airplane's exterior surface in a stationary environment, such as a maintenance, repair, and overhaul (MRO) shop [29]. Further an automated inspection that involves the utilization of 'stacked ensembles' of CNNs was implemented to classify defects found on aircraft surfaces. This approach utilized images obtained while inspecting the bores of aircraft propeller blades using a borescope. The development of a detection device focused on detecting air bubbles, impurities, and other flaws within the flexible connection of an aviation aircraft canopy was successfully accomplished [30]. By 2021, the focus shifted to artificial intelligence for automated defect detection systems and quality checks for the lining of aircraft interiors [31]. In 2022 a system that leverages precision instruments for intelligent inspection of subtle damages was developed [32].

For the challenging context of aircraft manufacturing, where the industrial environment is uncontrolled, a specialized single shot detector neural network (SSDNN) was developed to detect and measure drills and fixation elements like rivets and temporary fasteners [34]. The algorithm is optimized specifically for detecting circular elements, moving away from generic anchor boxes [34].

To identify surface anomalies on aircraft landing gear components, an anomaly detection strategy utilizing a convolutional autoencoder was developed [35]. This method facilitates the identification of defects during aircraft landing gear inspections. Additionally, the inspection of skin defects is an essential aspect of the traditional aircraft inspection process. Hence, an investigation was undertaken to explore a vision-based approach for pixel-level defect detection, employing the 'mask scoring R-CNN' framework. The experiment demonstrated that the novel classifier significantly improved the accuracy of detecting and segmenting aircraft skin defects, surpassing the performance of both the attention mechanism and feature fusion module [36].

A novel approach combines principles from natural language processing (NLP) and computer vision, integrating a knowledge graph to establish connections between welding appearance qualities and their corresponding locations in a point-wise space. To extract features within this space, a flexible graph convolution network (FGCN) algorithm is employed. For the challenging context of aircraft manufacturing, where the industrial environment is uncontrolled, a specialized single shot detector neural network (SSDNN) was developed to detect and measure drills and fixation elements like rivets and temporary fasteners [33]. The algorithm is optimized specifically for detecting circular elements, moving away from generic anchor boxes [33]. To identify surface anomalies on aircraft landing gear components, an anomaly detection strategy utilizing a convolutional autoencoder was developed [34]. A vision-based pixel-level defect detection approach using the 'mask scoring R-CNN' framework showed that the novel classifier significantly enhanced the accuracy of detecting and segmenting aircraft skin defects, outperforming attention mechanisms, and feature fusion modules [35]. A computer vision framework consisting of two steps was created to efficiently identify bolt loosening angles from UAV images. The method quantifies the bolt loosening angle using a number of computer vision technologies and uses a faster region-based convolutional neural network for bolt detection. By integrating the framework with a web server and an iOS app, field engineers can view inspection results quickly and increase the effectiveness of onsite maintenance [36]. The feasibility of employing small unmanned aerial vehicles (UAVs) for conducting visual inspections of aircraft structures was explored [37].

By 2023, augmented reality (AR) technologies were being combined with computer vision algorithms to assist users in inspection tasks, a novel method for detecting the twisting direction of safety wire in aviation fasteners using machine vision was proposed [38]. Simultaneously, safety wire status detection emerged as a significant area of interest, underlining the importance of computer vision in ensuring the safety of aircraft components. Automated systems capable of detecting the status of safety wires were developed, further emphasizing the role of computer vision in automating safety-critical tasks [39]. Automated inspection in smart manufacturing utilizes advanced technologies like computer vision and machine learning to enhance quality control. This transformative approach minimizes human intervention, optimizes inspection processes, and fosters a more adaptive manufacturing ecosystem, ensuring efficiency, accuracy, and proactive issue identification for improved overall production quality [40],[41]

8.5.1 Primary outcomes identified in notable papers within the literature review

Table 8.1 concisely compiles and presents the central insights and conclusions derived from a careful examination of significant papers featured in the comprehensive literature review. This table serves as a summary, capturing the essence of the key findings from prominent scholarly works, contributing to a comprehensive understanding of the subject matter under consideration.

8.5.2 Components, structure, and technological framework in computer vision

A detailed summary of the essential elements, structural framework, and technological aspects pertaining to computer vision for the inspection of aircraft and engines, as outlined by Padfield, Brooksby, and Kaucic in their 2007 [64] study, is meticulously presented in Table 8.2.

8.6 COMPUTER VISION FOR AERO ENGINE INSPECTION

The application of computer vision in aeroengine maintenance has seen significant advancements over the past few decades. In 1989, research focused on integrating automated systems, such as the Turbine Airfoil Laser Welding Work Center [65] at the Pratt & Whitney division of United Technologies Corporation, which improved productivity by 45%. By 1991, the automated combustor hole drilling and inspection system [66] had been developed, which combined laser drilling and machine vision inspection, potentially leading to substantial cost savings. In 2007, an 'aid to visual' [64] system capable of detecting minute post deformations in aeroengine high-pressure (HP) turbine disks was introduced, using algorithms such as edge detection, registration, and curve-similarity. A decade later, in 2017, a highly flexible slender continuum robot [64] was developed, which could navigate and perform complex trajectories in constrained environments like gas turbine engines. In 2019, the focus shifted towards algorithms for image inspection, with the introduction of a weighted morphology algorithm [67] for airplane bore scope crack detection. As of 2021, ongoing efforts were focused on the development of automated methods to detect defects on the edges of engine blades [68], with an accuracy rate of 83% achieved for a test size of 60 blades. In 2022, convolutional neural networks [69] were trained for crack tip detection in fatigue crack growth experiments, introducing a novel architecture, parallel nets. Most recently, in 2023, a innovative deep

Table 8.1 Key findings from literature review

Year	*Title*	*Methodology*	*Findings/ results*	*Ref.*
2019	"Vision-Based Aircraft Pose Estimation for UAVs Autonomous Inspection without Fiducial Markers"	Employing an RGB camera for autonomous aircraft detection, frames compared with a landmark database.	Demonstrated system effectiveness in real UAV flight scenarios, achieving precise pose localization. Proposed methods offer industrial advantages by removing constraints present in existing solutions.	[42]
2019	"Visual Inspection of the Aircraft Surface Using a Teleoperated Reconfigurable Climbing Robot and Enhanced Deep Learning Technique"	Robot that can be reconfigured ("Kiropter") captures aircraft surface images; SSD MobileNet framework enhances defect detection.	Experimental results confirm superior accuracy in detecting surface defects and surface stains when compared to conventional models.	[43]
2019	"Automating visual inspection with convolutional neural networks"	Surface defect detection using synthetically generated images and CNNs (U-Net).	The approach effectively detects anomalies, especially faults, in synthetic images of jets with varying angles and perspectives.	[44]
2020	"Investigation of Aircraft Surface Defects Detection Based on YOLO Neural Network"	YOLO neural network predicts skin crack, peeling, deformation, and tear with high accuracy.	YOLO neural network achieves prediction recognition accuracy surpassing faster-RCNN for skin crack (72%), peeling (84%), deformation (68%), and tear (70%).	[45]
2020	"Flexible Connection of Cockpit Canopy Defect Detection Device Based on Machine Vision"	Device uses OpenCV for intelligent detection of bonding surface defects.	The device improves efficiency, detecting flexible connection defects in cockpit canopies by approximately 50% compared to manual visual inspection.	[46]
2020	"A deep ensemble classifier for surface defect detection in aircraft visual inspection"	Aircraft surfaces with and without defects are categorized by a stacked ensemble of CNNs.	The proposed framework improves inspection accuracy during aircraft propeller blade borescope inspections, outperforming conventional single-model methods.	[47]

2020	"Using convolutional neural networks to automate aircraft maintenance visual inspection"	Hybrid method that combines augmentation methods with MASK R-CNN for defect identification.	Improved performance with an F1 score of 67.50% and an F2 score of 66.37% in detecting defects during aircraft maintenance visual inspection.	[48]
2021	"Automated defect detection and decision-support in gas turbine blade inspection"	Acquisition of images, creation of detection software, and development of a decision support tool.	Combined system detects and locates defects with an accuracy of 83%, quantifying morphological features of defect size and location.	[49]
2021	"Corrosion Detection on Aircraft Fuselage with Multi-teacher Knowledge Distillation"	CNN used for detecting corrosion spots on fuselage surface and rivets.	Proposed ensemble classifier significantly improves classification accuracy compared to the baseline single ResNet 50 neural network.	[50]
2021	"UAV-based inspection of airplane exterior screws with computer vision"	CNN was used to extract screws from photos and characterize interest zones.	Increased classification and detection performances, highlighting the importance of prior models during image analysis.	[51]
2022	"Anomaly detection for industrial surface inspection: application in maintenance of aircraft components"	For defect detection, an anomaly detection based on a convolutional autoencoder is used.	Results show successful reconstruction of normal and defective inputs, laying the foundation for machine vision-based surface inspection in aircraft maintenance.	[52]
2022	"A UAV-Based Aircraft Surface Defect Inspection System via External Constraints and Deep Learning"	UAV-based defect detection using a high-resolution camera, implemented based on deep learning.	Experiments in simulation and real data environments demonstrate the effectiveness of the proposed external localization module and crack detection module.	[53]

(continued)

Table 8.1 (*Cont.*)

Year	*Title*	*Methodology*	*Findings/ results*	*Ref.*
2022	"UAS based autonomous visual inspection of airplane surface defects"	Utilizing a UAS for data collection, with a 3D-LiDAR based SLAM algorithm for position and orientation data.	Reduces airplane ground time, increasing efficiency in the inspection process.	[54]
2022	"Visual inspection of aircraft skin: Automated pixel-level defect detection by instance segmentation"	Introducing attention mechanism, feature fusion module, and a new classifier head for improved feature representation and segmentation precision.	Proposed method outperforms mask R-CNN and mask scoring R-CNN in pixel-level aircraft skin defect detection, increasing segmentation precision.	[55]
2022	"Gap Detection of Aircraft Bell-Shaped Tube Based on Convolutional Neural Networks"	Improved YOLOv4 algorithm for gap detection in aircraft bell-shaped tubes.	Enhanced algorithm achieves higher average precision (92.21%) without compromising detection speed compared to the original YOLOv4.	[56]
2023	"Surface defect detection algorithm based on improved YOLOv4"	On the test set, improvements to YOLOv4 result in a mean average precision (mAP) of 90.10%.	Improved YOLOv4 achieves a higher mAP and detection speed compared to traditional YOLOv4.	[57]
2023	"Surface Defect Detection of Aeroengine Blades Based on Cross-Layer Semantic Guidance"	Cross-layer semantic guided network (CSGNet) based on YOLOv6 for defect detection.	Detection results show increased accuracy with fewer parameters compared to YOLOv6.	[58]
2023	"An Autonomous Aircraft Inspection System using Collaborative Unmanned Aerial Vehicles"	Details on methodology not disclosed.	Adaptable visual inspection system for aircraft using collaborative autonomous UAVs with minimal manual intervention.	[59]
2023	"AI-based Endpoint Detection for Surface Defect Removal on Aircraft Components"	YOLO v4 object detection network trained on real and synthetic data for endpoint detection.	System effectively monitors defect removal on landing gear components, providing feedback when the removal endpoint is achieved.	[60]

2023	"Drone-Based Inspection of the Appearance Defects for a Large Object"	Drone with mounted camera inspects large aircraft exteriors; YOLOv4-P6 network for defect recognition.	Drone-based system successfully detects defects like paint loss, corrosion, and dents on large aircraft exteriors.	[61]
2023	"Development of Image Processing Techniques in Crack Detection and Analysis"	UAV-based inspection system for detecting crack locations on aircraft's outer skin, further investigated with thermal and ultrasound imaging methods.	Reviews image processing algorithms designed for digital image processing on aircraft fuselage surface images.	[62]
2023	"Deep Learning-Based Real-Time Defect Detection for Optimization of Aircraft Manufacturing and Control Performance"	Real-time deep learning-based system for detecting faulty components at various stages of production.	Convolutional neural network (CNN) used for recognizing abnormal states. Results show significant reduction in time delays and total costs, improving operational cost management in the aerospace industry	[63]

Table 8.2 Key components, architecture, and technologies in computer vision

Components	*Architecture*	*Technologies*
Cameras or sensors	Data acquisition layer: capture visual data	Machine learning and deep learning
Image processing unit	Preprocessing layer: enhance and prepare data	Computer vision libraries (e.g., OpenCV)
Feature extraction module	Feature extraction and analysis layer: identify features and analyze data	Neural networks (e.g., CNNs)
Machine learning algorithms	Not available	IoT (Internet of Things)
Pattern recognition system	Not available	Cloud computing
Decision-making module	Decision-making layer: process analyzed data and make decisions	Not available
	User interface layer: provide a user-friendly interface for interaction	Not available

learning framework [70] was developed for automated borescope inspection, achieving precisions and recalls of over 90%. Moreover, the inspection of aeroengine blades represented a critical application of computer vision in the aerospace industry. Researchers conducted a thorough review of defect detection methods in aeroengine blades, highlighting the role of computer vision in maintaining the operational integrity of these crucial components [71]. These technologies can also be integrated in smart manufacturing domain [40].

8.7 USE CASES FOR THE APPLICATION OF COMPUTER VISION (CV) IN THE INSPECTION OF AIRCRAFT AND AEROENGINES

Prominent use cases for application of computer vision (CV) for inspection of aircraft and aeroengine have been complied in Table 8.3. Although many of these use cases are yet to be offered as a commercial solution, they hold a potential in mitigating human error in aircraft and engine inspection.

8.8 LIMITATIONS AND FUTURE WORK

Computer vision applications in aircraft and engine/gas turbine inspection offer promising potential for enhancing safety and operational efficiency in the aviation industry. However, there are challenges that need to be addressed. Robust algorithms are required to handle diverse visual

Table 8.3 Important use cases for application of computer vision in aircraft and aeroengine inspection

Use cases	*Description*	*Ref.*
Detection of aircraft dents	Combining MASK R-CNN and various augmentation techniques, the improved approach achieves higher accuracy in detecting aircraft dents and other defects.	[72] [73]
Detection of lightning strike damage	Automated inspection system using multi-copter to locate and identify lightning strike-induced defects like pits, burn marks, and holes on aircraft surfaces through digitalization and automatic image evaluation using industrial camera.	[74]
Screw inspection for both loose and tight conditions	An UAV with **LIDAR** locates itself in an aircraft frame, using CNN to extract screws from images. Classic matching techniques improve screw recognition and detect missing ones using computer vision algorithms.	[27]
Inspection of paint quality	Computer vision detects paint defects like flaking, wear, and degradation, reducing human errors. KLM engineering & maintenance has collaborated with Mainblades, a startup, to implement drone inspection technology to address the long-standing paint peeling issue experienced by operators of Boeing 787 and Airbus A350 aircraft.	[35] [75]
Inspection of aircraft screws	A UAV with lidar technology captures precise images of aircraft exterior screws. A CNN-based method detects screws and matches them with a prior model, identifying missing screws. For streamlined aircraft maintenance, computer vision algorithms are employed to assess the condition of each screw, effectively detecting any missing or loose ones.	[27]
Visual recognition of aircraft mechanical parts	A visual module has been developed with the enhanced ability to identify and precisely locate diverse mechanical components within a fighter aircraft. specifically, the Alenia-Aermacchi M346. The assessment of this module involved analyzing a set of 248 images, encompassing eight unique panels and 20 diverse mechanical components. These images were captured under diverse imaging conditions, ensuring a comprehensive assessment of the module's performance.	[76]

(continued)

Table 8.3 (*Cont.*)

Use cases	*Description*	*Ref.*
Aeroengine blade	The existing software solution employs computer vision and image processing methods to identify defects on the leading edges and trailing edges of aeroengine compressor blades. Even with a limited dataset, the software achieves detection performance on par with that of human inspectors and neural networks.	[68]
Aeroengine fan blade inspection for nicks and dents during engine production	AV&R's OwlVision software specialized in detecting and analyzing surface defects on fan blades, including leading edge defects and dynamic inspection capabilities for precise 3D measurements of critical areas requiring validation.	[77]
Detecting damage on aeroengine rotor blades through borescope videos inputs.	An automatic workflow using deep learning to detect damage on rotor blades from borescope videos, enabling fast and accurate sentencing.	[70] [78]

data, including varying lighting conditions, perspectives, and occlusions, to improve defect detection accuracy. Real-time implementation of computer vision requires consideration of computational power and memory resources [79] for seamless integration into maintenance scenarios [80]. Domain gap impacts the performance of models trained purely on synthetic data when tested on real data [81]. Defect detection in airplane fuselage encounters a significant issue related to washed or unwashed surfaces. Detecting defects on unwashed fuselage, which may contain dirt and debris, poses challenges in the inspection process [82]. Approval from regulatory authorities and original equipment manufacturers (OEMs) poses unique challenges that must be diligently handled. Meeting certification standards set by bodies like FAA and EASA is crucial for aviation regulation compliance [83, 84]. Comprehensive validation and testing are crucial to establish the dependability and precision of computer vision systems across diverse real-world situations. Ensuring interoperability with existing infrastructure and addressing human–machine interaction are critical for integrating computer vision into maintenance workflows. Robustness in adverse conditions, long-term reliability, and cost-effectiveness are also vital considerations.

8.9 CONCLUSION

This chapter introduces a transformative approach to aircraft exterior inspection using advanced computer vision technology. The chapter highlights the challenges faced by aircraft technicians during routine inspection procedures and the potential for human error due to the repetitive nature and complexity involved in analyzing intricate structures. By automating the inspection process, this methodology aims to mitigate the risk of overlooking defects or abnormalities that may have serious implications on aircraft safety. The chapter covers major use cases and discusses their significance in enhancing the inspection process for aircraft and engines. Currently, most computer vision solutions for aircraft and engine inspection are in the technical proof-of-concept stages. Full-scale deployment requires further refinement, validation, and regulatory approvals. As the technology matures and overcomes limitations, these applications can revolutionize aircraft maintenance and inspection processes. The chapter also acknowledges the challenges faced in deploying computer vision solutions for aircraft and engine inspection, including the need for further refinement, validation, and regulatory approvals. The implementation of computer vision techniques in the aviation industry will require careful consideration and gradual adoption. Continuing research and development are essential to adapt computer vision techniques to different aircraft and engine models and types, enabling broader applicability of the technology. Overall, the chapter highlights the potential of computer vision technology in optimizing safety measures and reducing errors during routine inspection, contributing to a safer aviation industry. Currently, most computer vision solutions for aircraft and engine inspection are in the technical proof-of-concept stages. Full-scale deployment requires further refinement, validation, and regulatory approvals. As the technology matures and overcomes limitations, these applications can revolutionize aircraft maintenance and inspection processes. However, their implementation in the aviation industry will require careful consideration and gradual adoption. Continuing research and development are essential to adapt computer vision techniques to different aircraft and engine models and types, enabling broader applicability of the technology. In summary, this work highlights the immediate impact of the methodology on improving safety measures and reducing errors in routine aircraft inspection, with a forward-looking perspective on the transformative potential of computer vision technology in the aviation industry

REFERENCES

[1] B. S. Dhillon, "Human reliability and error in transportation systems." in *Springer series in reliability engineering*. London: Springer, 2007.

[2] I. L. Davis and M. Siegel, "Automated nondestructive inspector of aging aircraft," in *Measurement Technology and Intelligent Instruments*, SPIE, pp. 190–201, Sep. 1993. doi: 10.1117/12.156432.

[3] M. Siegel, P. Gunatilake, and G. Podnar, "Robotic assistants for aircraft inspectors," *IEEE Instrumentation & Measurement Magazine*, vol. 1, no. 1, pp. 16–30, Mar. 1998. doi: 10.1109/5289.658190.

[4] P. Backes, Y. Bar-Cohen, and B. Joffe, "The multifunction automated crawling system (MACS)," vol. 1, pp. 335–340, May 1997, doi: 10.1109/ROBOT.1997.620060.

[5] M. Futterlieb, J. Frejaville, F. Donadio, M. Devy, and S. Larnier, "Air-cobot: aircraft enhanced inspection by smart and collaborative robot," 2017. Accessed: Jul. 19, 2023. [Online]. Available: www.semanticscholar.org/paper/Air-Cobot-%3A-Aircraft-Enhanced-Inspection-by-Smart-Futterlieb-Frejaville/0a4c4c9a56465b700e4e0d68f683b8a344cd02f9.

[6] M. Futterlieb, "Vision based navigation in a dynamic environment," PhD Thesis, Université Paul Sabatier – Toulouse III, 2017. Accessed: Jul. 19, 2023. [Online]. Available: https://hal.laas.fr/tel-01624233.

[7] I. Jovančević, "Exterior inspection of an aircraft using a Pan-Tilt-Zoom camera and a 3D scanner moved by a mobile robot: 2D image processing and 3D point cloud analysis," PhD Thesis , Ecole des Mines d'Albi-Carmaux, 2016. Accessed: Jul. 19, 2023. [Online]. Available: https://theses.hal.science/tel-01687831.

[8] M.-A. Bauda, A. Grenwelge, and S. Larnier, "3D scanner positioning for aircraft surface inspection," in *9th European Congress on Embedded Real Time Software and Systems (ERTS 2018)*. Toulouse, France, Jan. 2018. Accessed: Jul. 19, 2023. [Online]. Available: https://hal.science/hal-02156494.

[9] M. Rice, *Automating the Visual Inspection of Aircraft*, Institute for Infocomm Research, 2018. Accessed: Jul. 19, 2023. [Online]. Available: https://oar.a-star.edu.sg/communities-collections/articles/13872.

[10] Y. D. V. Yasuda, F. A. M. Cappabianco, L. E. G. Martins, and J. A. B. Gripp, "Aircraft visual inspection: a systematic literature review," *Computers in Industry*, vol. 141, p. 103695, Oct. 2022, doi: 10.1016/j.compind.2022.103695.

[11] V. Wiley and T. Lucas, "Computer vision and image processing: a paper review," *International Journal of Artificial Intelligence Research*, vol. 2, p. 22, Feb. 2018, doi: 10.29099/ijair.v2i1.42.

[12] J. Vora, S. Nair, A. K. Gramopadhye, A. T. Duchowski, B. J. Melloy, and B. Kanki, "Using virtual reality technology for aircraft visual inspection training: presence and comparison studies," *Applied Ergonomics*, vol. 33, no. 6, pp. 559–570, Nov. 2002, doi: 10.1016/S0003-6870(02)00039-X.

[13] M. Siegel and P. Gunatilake, "Enhanced remote visual inspection of aircraft skin." In *Proceedings of the Intelligent NDF Sciences for Aging and Futuristic Aircraft Workshops*, pp. 101–112. 1997.

[14] "*Visual Inspection for Aircraft.Advisory Circular ACNO. 43-204.*" Federal Aviation Administration, 1997.

[15] C. G. Drury and J. Watson, "Good practices in visual inspection." Federal Aviation Administration, Flight Standards Service, 2002. [Online]. Available: www.faa.gov/about/office_org/headquarters_offices/avs/offices/afx.

[16] K. A. Latorella and P. V. Prabhu, "A review of human error in aviation maintenance and inspection," *International Journal of Industrial Ergonomics*, vol. 26, pp. 521–549, 2000.

[17] J. C. Pereira and J. C. P. Souza, "Prevention of human errors in the visual inspection during overhaul of aeroengines – a case study," *Presented at the ICIEOM –28 International Conference on Industrial Engineering and Operations Management*, 2022. [Online]. Available: http://portalabepro.educacao.ws/ijcieom/restrito/arquivos/icieom2022/FULL_0023_37514.pdf.

[18] R. C. Movich, "Robotic drilling and riveting using computer vision," *SME Tech Pap Ser MS*, 1980 [Online]. Available: www.scopus.com/inward/record.uri?eid=2-s2.0-85040272062&partnerID=40&md5=b886116e6ccd4637127d111e25c59860.

[19] O. R. Mitchell and T. A. Grogan, "Global and partial shape discrimination for computer vision," *Optical Engineering*, vol. 23, no. 5, pp. 484–491, 1984, doi: 10.1117/12.7973326.

[20] A. Rosenfeld, "Computer vision," *Advanced Computers*, vol. 27, no. C, pp. 265–308, 1988, doi: 10.1016/S0065-2458(08)60261-2.

[21] P. Freeman, "A robust method of countersink inspection using machine vision," in *SAE Techncial Paper.*, SAE International, 2004. doi: 10.4271/2004-01-2820.

[22] B. Li, X. Wang, H. Yang, and Z. Zhou, "Aircraft rivets defect recognition method based on magneto-optical images," in *Int. Conf. Mach. Vis. Hum.-Mach. Interface, MVHI*, 2010, pp. 788–791. doi: 10.1109/MVHI.2010.52.

[23] M. Dixon *et al.*, "Measuring optical distortion in aircraft transparencies: a fully automated system for quantitative evaluation," *Machine Vision Applications*, vol. 22, no. 5, pp. 791–804, 2011, doi: 10.1007/s00138-010-0258-z.

[24] I. Jovančević, J.-J. Orteu, T. Sentenac, and R. Gilblas, "Automated visual inspection of an airplane exterior," in *Proc SPIE Int Soc Opt Eng*, Meriaudeau F. and Aubreton O., Eds., SPIE, 2015. doi: 10.1117/12.2182811.

[25] I. Jovančević, A. Arafat, J.-J. Orteu, and T. Sentenac, "Airplane tire inspection by image processing techniques," in *5th Mediterranean Conference on Embedded Computing (MECO'2016)*, Bar, Montenegro, Jun. 2016. doi: 10.1109/MECO.2016.7525733.

[26] D. A. Luo, E. T. Barraza, and M. W. Kudenov, "Aircraft skin defect localization using imaging polarimetry," *Optical Engineering*, vol. 57, no. 8, 2018, doi: 10.1117/1.OE.57.8.084101.

[27] J. Miranda, S. Larnier, A. Herbulot, and M. Devy, "UAV-based inspection of airplane exterior screws with computer vision." In *14th International Joint Conference on Computer Vision, Imaging and Computer Graphics Theory and Applications*. 2019.

[28] B. Lawless, R. Heinemann, and A. Liaqat, "Assessing the quality of inspection for tapered aircraft fastener holes using an engineer's blue contact test," *Precision Engineering*, vol. 63, pp. 62–67, 2020, doi: 10.1016/j.precisioneng.2020.01.004.

[29] Y. Sun and O. Ma, "Automating aircraft exterior inspection with robotics and machine learning technologies," in *ASME Int Mech Eng Congress Expos Proc*, American Society of Mechanical Engineers (ASME), 2020. doi: 10.1115/IMECE2020-24577.

[30] P. Wang, X. Lu, J. Chen, and P. Zhang, "Flexible connection of cockpit canopy defect detection device based on machine vision," in *IOP Conf. Ser. Mater. Sci. Eng.*, Institute of Physics Publishing, 2020. doi: 10.1088/1757-899X/711/1/012007.

[31] N. Mosca, V. Renò, M. Nitti, C. Patruno, and E. Stella, "Post assembly quality inspection using multimodal sensing in aircraft manufacturing," in *Proc SPIE Int Soc Opt Eng*, Stella E., Ed., SPIE, 2021. doi: 10.1117/12.2594104.

[32] J. Tao, Q. Chen, J. Xu, H. Zhao, and S. Song, "Utilization of both machine vision and robotics technologies in assisting quality inspection and testing," *Mathematical Problem Engineering*, vol. 2022, 2022, doi: 10.1155/2022/7547801.

[33] L. Ruiz, S. Díaz, J. M. González, and F. Cavas, "Improvement of fixation elements detection in aircraft manufacturing," in *Lect. Notes Comput. Sci.*, Ferrández Vicente J.M., Álvarez-Sánchez J.R., de la Paz López F., and Adeli H., Eds., Springer Science and Business Media Deutschland GmbH, 2022, pp. 374–382. doi: 10.1007/978-3-031-06527-9_37.

[34] F. Kähler, O. Schmedemann, and T. Schüppstuhl, "Anomaly detection for industrial surface inspection: application in maintenance of aircraft components," in *Procedia CIRP*, Valente A., Carpanzano E., and Boer C., Eds., Elsevier B.V., 2022, pp. 246–251. doi: 10.1016/j.procir.2022.05.197.

[35] M. Ding, B. Wu, J. Xu, A. N. Kasule, and H. Zuo, "Visual inspection of aircraft skin: automated pixel-level defect detection by instance segmentation," *Chinese Journal of Aeronautics*, 2022, doi: 10.1016/j.cja.2022.05.002.

[36] Y. Qi, P. Li, B. Xiong, S. Wang, C. Yuan, and Q. Kong, "A two-step computer vision-based framework for bolt loosening detection and its implementation on a smartphone application," *Structural Health Monitoring*, vol. 21, no. 5, pp. 2048–2062, 2022, doi: 10.1177/14759217211049995.

[37] G. Y. Pardomoan, A. Asyrof, and Y. I. Jenie, "Micro aerial vehicle as an aircraft inspection system based on computer vision," in *AIAA AVIATION Forum*, American Institute of Aeronautics and Astronautics Inc, AIAA, 2022. doi: 10.2514/6.2022-3406.

[38] Z. Zhang, H. Wang, Z. Pan, W. Li, and Z. Su, "Increasing trust with augmented reality in human-in-the-loop intelligent system: a case study with inspection task," in *Proc. – IEEE Conf. Virtual Real. 3D User Interfaces Abstr. Workshops, VRW*, Institute of Electrical and Electronics Engineers Inc., 2023, pp. 743–744. doi: 10.1109/VRW58643.2023.00214.

[39] J. Liu, S. Zhao, and F. Hu, "Automatic detection of fastener safety wire twisting direction based on machine vision," in *Proc SPIE Int Soc Opt Eng*, Zhang T. and Yang T., Eds., SPIE, 2023. doi: 10.1117/12.2660005.

[40] A. AlMangour Hari Singh, Parveen, Bandar, Ed., *Handbook of Smart Manufacturing: Forecasting the Future of Industry 4.0*. Boca Raton: CRC Press, 2023. doi: 10.1201/9781003333760.

[41] S. Rani, K. Tripathi, and A. Kumar, "Machine learning aided malware detection for secure and smart manufacturing: a comprehensive analysis of the state of the art," *International Journal on Interactive Design and Manufacturing (IJIDeM)*, Oct. 2023, doi: 10.1007/s12008-023-01578-0.

[42] D. Cazzato, M. A. Olivares-Mendez, J. L. Sanchez-Lopez, and H. Voos, "Vision-based aircraft pose estimation for UAVs autonomous inspection without fiducial markers," in *IECON 2019 – 45th Annual Conference of the IEEE Industrial Electronics Society*, Oct. 2019, pp. 5642–5648. doi: 10.1109/IECON.2019.8926667.

[43] B. Ramalingam *et al.*, "Visual inspection of the aircraft surface using a teleoperated reconfigurable climbing robot and enhanced deep learning technique," *International Journal of Aerospace Engineering*, vol. 2019, p. e5137139, Sep. 2019, doi: 10.1155/2019/5137139.

[44] C. D. Hollander, S. Das, and S. Suliman, "Automating visual inspection with convolutional neural networks," *Presented at the Proceedings of the Annual Conference of the Prognostics and Health Management Society*, PHM, 2019. doi: 10.36001/phmconf.2019.v11i1.868.

[45] Z. Deyin, W. Penghui, T. Mingwei, C. Conghan, W. Li, and H. Wenxuan, "Investigation of aircraft surface defects detection based on YOLO neural network," in *2020 7th International Conference on Information Science and Control Engineering (ICISCE)*, Dec. 2020, pp. 781–785. doi: 10.1109/ICISCE50968.2020.00165.

[46] P. Wang, X. Lu, J. Chen, and P. Zhang, "Flexible connection of cockpit canopy defect detection device based on machine vision," *IOP Conference Series: Materials Science and Engineering*, vol. 711, p. 012007, Jan. 2020, doi: 10.1088/1757-899X/711/1/012007.

[47] I. Ren, F. Zahiri, G. Sutton, T. Kurfess, and C. Saldana, "A deep ensemble classifier for surface defect detection in aircraft visual inspection," *Smart Sustainable Manufacture System*, vol. 4, no. 1, p. 20200031, Sep. 2020, doi: 10.1520/SSMS20200031.

[48] A. Doğru, S. Bouarfa, R. Arizar, and R. Aydoğan, "Using convolutional neural networks to automate aircraft maintenance visual inspection," *Aerospace*, vol. 7, no. 12, Art. no. 12, Dec. 2020, doi: 10.3390/aerospace7120171.

[49] J. Aust, S. Shankland, D. Pons, R. Mukundan, and A. Mitrovic, "Automated defect detection and decision-support in gas turbine blade inspection," *Aerospace*, vol. 8, no. 2, pp. 1–27, 2021, doi: 10.3390/aerospace8020030.

[50] K. Zuchniak, W. Dzwinel, E. Majerz, A. Pasternak, and K. Dragan, "Corrosion detection on aircraft fuselage with multi-teacher knowledge distillation," *Lecture Notes in Computer Science (Including Subseries Lecture Notes in Artificial Intelligence and Lecture Notes in Bioinformatics)*, vol. 12747 LNCS, pp. 318–332, 2021, doi: 10.1007/978-3-030-77980-1_25.

[51] J. Miranda, S. Larnier, A. Herbulot, and M. Devy, "UAV-based inspection of airplane exterior screws with computer vision." In *14th International Joint Conference on Computer Vision, Imaging and Computer Graphics Theory and Applications*. 2019.

[52] F. Kähler, O. Schmedemann, and T. Schüppstuhl, "Anomaly detection for industrial surface inspection: application in maintenance of aircraft components," *Procedia CIRP*, vol. 107, pp. 246–251, 2022, doi: 10.1016/j.procir.2022.05.197.

[53] Y. Liu, J. Dong, Y. Li, X. Gong, and J. Wang, "A UAV-based aircraft surface defect inspection system via external constraints and deep learning," *IEEE*

Transactions on Instrumentation and Measurement, vol. 71, pp. 1–1, Jan. 2022, doi: 10.1109/TIM.2022.3198713.

[54] M. Tappe *et al.*, *UAS Based Autonomous Visual Inspection of Airplane Surface Defects*, 2022. doi: 10.1117/12.2612579.

[55] M. Ding, B. Wu, J. Xu, A. N. Kasule, and H. Zuo, "Visual inspection of aircraft skin: automated pixel-level defect detection by instance segmentation," *Chinese Journal of Aeronautics*, 2022, doi: 10.1016/j.cja.2022.05.002.

[56] Y. Yao, F. Li, and J. Wang, "Gap detection of aircraft bell-shaped tube based on convolutional neural networks," *Lecture Notes on Data Engineering and Communications Technologies*, vol. 89, pp. 1340–1349, 2022, doi: 10.1007/978-3-030-89698-0_138.

[57] Li B, Wang C, Ding X Y, et al., "Surface defect detection algorithm based on improved YOLOv4," *Journal of Beijing University of Aeronautics and Astronautics*, no. 49, pp. 710–717, 2023, Accessed: Nov. 11, 2023. [Online]. Available: https://bhxb.buaa.edu.cn/bhzk/en/article/doi/10.13700/j.bh.1001-5965.2021.0301

[58] Z. Deyin, W. Penghui, T. Mingwei, C. Conghan, W. Li, and H. Wenxuan, "Investigation of aircraft surface defects detection based on YOLO neural network," in *2020 7th International Conference on Information Science and Control Engineering (ICISCE)*, Dec. 2020, pp. 781–785. doi: 10.1109/ICISCE50968.2020.00165.

[59] A. Saha, L. Kumar, S. Sortee, and B. C. Dhara, "An autonomous aircraft inspection system using collaborative unmanned aerial vehicles," in *2023 IEEE Aerospace Conference*, Mar. 2023, pp. 1–10. doi: 10.1109/AERO55745.2023.10115655.

[60] F. Kähler, A. C. K. Shetty, and T. Schüppstuhl, "AI-based endpoint detection for surface defect removal on aircraft components," in *2023 IEEE/SICE International Symposium on System Integration (SII)*, Jan. 2023, pp. 1–6. doi: 10.1109/SII55687.2023.10039239.

[61] W.-J. Wang, X.-Y. Dai, C.-Y. Cheng, and S.-M. Ciou, "Drone-based inspection of the appearance defects for a large object," in *2023 International Conference on System Science and Engineering (ICSSE)*, Jul. 2023, pp. 1–5. doi: 10.1109/ICSSE58758.2023.10227178.

[62] L. S. Al Dhafari, A. Afzal, R. I. Al Bahrani, Y. Al Busaidi, A. Sheikh-Akbari, and M. S. Hossain, "Development of image processing techniques in crack detection and analysis," in *2023 3rd International Conference on Electrical, Computer, Communications and Mechatronics Engineering (ICECCME)*, Jul. 2023, pp. 1–7. doi: 10.1109/ICECCME57830.2023.10252690.

[63] I. Shafi *et al.*, "Deep learning-based real time defect detection for optimization of aircraft manufacturing and control performance," *Drones*, vol. 7, no. 1, 2023, doi: 10.3390/drones7010031.

[64] D. Padfield, G. Brooksby, and R. Kaucic, "Automatic deformation detection for aircraft engine disk inspection," *Journal of Multimedia*, vol. 2, no. 4, pp. 34–44, 2007, doi: 10.4304/jmm.2.4.34-44.

[65] B. E. Gardner and J. M. Bartlett, "Automated laser welding of turbine airfoil details," in *Proceedings of ASME Turbo Expo*, American Society of Mechanical Engineers (ASME), 1989. doi: 10.1115/89-GT-327.

[66] F. R. Moskal, "Automated combustor hole drilling and inspection system," in *Annu Forum Proc Am Helicopter Soc*, Publ by American Helicopter Soc, 1991, pp. 391–398. [Online]. Available: www.scopus.com/inward/record.uri?eid=2-s2.0-0026153661&partnerID=40&md5=d878f1a89624bb8b0c98d52a10d8f527.
[67] Y.-Y. Guo, Z.-G. Liu, and Q.-J. Gao, "Based on weighted morphology aero-engine bore scope cracks image segmentation," in *ACM Int. Conf. Proc. Ser.*, Association for Computing Machinery, 2019, pp. 451–455. doi: 10.1145/3318299.3318371.
[68] J. Aust, S. Shankland, D. Pons, R. Mukundan, and A. Mitrovic, "Automated defect detection and decision-support in gas turbine blade inspection," *Aerospace*, vol. 8, no. 2, pp. 1–27, 2021, doi: 10.3390/aerospace8020030.
[69] D. Melching, T. Strohmann, G. Requena, and E. Breitbarth, "Explainable machine learning for precise fatigue crack tip detection," *Science Report*, vol. 12, no. 1, 2022, doi: 10.1038/s41598-022-13275-1.
[70] A. Upadhyay, J. Li, S. King, and S. Addepalli, "A deep-learning-based approach for aircraft engine defect detection," *Machine*, vol. 11, no. 2, 2023, doi: 10.3390/machines11020192.
[71] Y. Abdulrahman, M. A. Mohammed Eltoum, A. Ayyad, B. Moyo, and Y. Zweiri, "Aero-engine blade defect detection: a systematic review of deep learning models," *IEEE Access*, vol. 11, pp. 53048–53061, 2023, doi: 10.1109/ACCESS.2023.3280992.
[72] M. Sprehe, "AI meets drones – detecting damages on aircraft with computer vision," *Mainblades*. Accessed: Jul. 23, 2023. [Online]. Available: www.mainblades.com/blog-posts/ai-meets-drones-detecting-damages-on-aircraft-with-computer-vision
[73] A. Doğru, S. Bouarfa, R. Arizar, and R. Aydoğan, "Using convolutional neural networks to automate aircraft maintenance visual inspection," *Aerospace*, vol. 7, no. 12, Art. no. 12, Dec. 2020, doi: 10.3390/aerospace7120171.
[74] S. Bouarfa, A. Doğru, R. Arizar, R. Aydogan, and J. Serafico, "Towards automated aircraft maintenance inspection. a use case of detecting aircraft dents using Mask R-CNN," Jan. 2020. doi: 10.2514/6.2020-0389.
[75] N. P. Avdelidis, A. Tsourdos, P. Lafiosca, R. Plaster, A. Plaster, and M. Droznika, "Defects recognition algorithm development from visual UAV inspections," *Sensors*, vol. 22, no. 13, Art. no. 13, Jan. 2022, doi: 10.3390/s22134682.
[76] C. Cusano and P. Napoletano, "Visual recognition of aircraft mechanical parts for smart maintenance," *Computers in Industry*, vol. 86, pp. 26–33, Apr. 2017, doi: 10.1016/j.compind.2017.01.001.
[77] marlene, "Fan Blade Inspection," AV&R. Accessed: Jul. 24, 2023. [Online]. Available: https://avr-global.com/technical-articles/fan-blade-inspection/.
[78] C. Y. Wong, P. Seshadri, and G. T. Parks, "Automatic borescope damage assessments for gas turbine blades via deep learning," in *AIAA Scitech 2021 Forum*, Jan. 2021. doi: 10.2514/6.2021-1488.
[79] N. O'Mahony *et al.*, "Deep learning vs. traditional computer vision," in *Advances in Computer Vision*, K. Arai and S. Kapoor, Eds., in Advances in

Intelligent Systems and Computing, Vol. 943., Cham: Springer International Publishing, 2020, pp. 128–144. doi: 10.1007/978-3-030-17795-9_10.

[80] X. Feng, Y. Jiang, X. Yang, M. Du, and X. Li, "Computer vision algorithms and hardware implementations: a survey," *Integration*, vol. 69, pp. 309–320, Nov. 2019, doi: 10.1016/j.vlsi.2019.07.005.

[81] K. Man and J. Chahl, "A review of synthetic image data and its use in computer vision," *Journal of Imaging*, vol. 8, no. 11, p. 310, Nov. 2022, doi: 10.3390/jimaging8110310.

[82] T. Malekzadeh, M. Abdollahzadeh, H. Nejati, and N.-M. Cheung, "Aircraft fuselage defect detection using deep neural networks," *Presented at the 5th IEEE Global Conference on Signal and Information Processing*, Montreal, Canada, 2017.

[83] Y. Jiang, T. H. Tran, and L. Williams, "Machine learning and mixed reality for smart aviation: applications and challenges," *Journal of Air Transport Management*, vol. 111, p. 102437, Aug. 2023, doi: 10.1016/j.jairtraman.2023.102437.

[84] F. Boniol *et al.*, *Challenges in the Certification of Computer Vision-Based Systems, Aerospace Lab*, Sep. 2020, doi: 10.12762/2020.AL15-06.

Chapter 9

Unleashing the power of YOLOv5

Revolutionizing person detection and counting in restricted zones

Suguna R.,[1] Suriya Prakash J.,[2] and Neethu P.S.[3]

[1]Department of Computer Science and Engineering, Vel Tech Rangarajan Dr. Sagunthala R&D Institute of Science and Technology, Avadi, Chennai, India
[2]Department of Computer Science and Engineering, JAIN (Deemed-to-be-University), Bangaluru, India
[3]Department of Electronics and Communication Engineering, School of Engineering and Technology, Christ (Deemed-to-be-University), Bangaluru, India

9.1 INTRODUCTION

Artificial intelligence is a concept used in computer science to describe objects that can recognize situations and take appropriate action in response. The use of artificial intelligence has changed society and the economy [1]. Industries and organizations can find the information they need to implement smart manufacturing towards sustainable manufacturing practices [2, 3]. In the business world, managerial choices greatly influence manufacturing success and how manufacturing processes deal with unpredictability and uncontrollable aspects brought by globalization [4]. Typical attack approaches are evident in smart factories due to open standards and the internet's opening [5]. Machine learning algorithms are built to keep up with the complexity and speed of malware development found to be more reliable than traditional systems [6]. Machine learning methods also work well in cloud computing security attack detection [7].

The entire AI sector has revolutionized since the advent of artificial neural networks. Within the field of artificial intelligence, deep learning is a sophisticated method that produces reliable outcomes. Complex, industry-relevant challenges can be solved with the aid of deep learning. One such application is crowd counting. A method to determine the approximate number of people in an image is called 'crowd counting.' Determining the precise count of individuals or items in a single image is a difficult but important undertaking that finds use in various fields, including public safety and urban planning. Because crowd counting has a special significance for social security and development, it is especially common in various object-counting activities.

For various applications, including crowd management and public safety, accurate detection and counting of persons within specific zones is essential

DOI: 10.1201/9781003502470-9

[8]. Traditional person detection systems frequently struggle to produce real-time and precise findings in dynamic situations with shifting illumination, occlusions, and scale. Object detection frameworks using deep learning have transformed the discipline, considerably increasing both speed and accuracy [9].

The COVID-19 pandemic demonstrated that physical separation between infected individuals is one of the most crucial factors in virus transmission. The World Health Organization (WHO) mandates that only few people be present in indoor areas. The maximum number of people who can fit into a space varies depending on its size. Computers can be used to ensure that the capacity rule is correctly applied indoors under video surveillance [10]. An appropriate technique is essential for measuring the size of a predetermined zone in the video and counting the persons there in real time [11].

Traditional crowd-counting techniques like manual registers and sensors no longer work in today's dynamic, unpredictable environments. For monitoring individuals in different places, video surveillance has become essential. Many cutting-edge algorithms, including median flow and deep learning-based kernelized filters, have been used for precise people counting and real-time object tracking [12].

Person detection algorithms, such as sliding window techniques or selective search, are computationally expensive and struggle to offer real-time results. YOLOv5, on the other hand, uses a single-shot detection paradigm, allowing them to reach exceptional speed without sacrificing accuracy. These models employ convolutional neural networks (CNNs) to estimate bounding box coordinates and class probabilities for several objects in an image simultaneously [13].

The study is focused on customizing YOLOv5 to handle human identification and counting within predetermined zones. In crowd surveillance applications, such zones could represent restricted areas, entrances, or regions of interest. It is aimed to provide effective crowd management, early detection of security risks, and informed decision-making by correctly detecting and counting people within these zones [14].

The findings have comprehensive implications. The accurate detection and counting of people in restricted zones enable security officers, event organizers, and government agencies to manage crowds more efficiently, identifying potential security problems, and optimizing resource allocation. The models' real-time insights enable proactive decision-making, improving public safety and operational efficiency.

The limitations of previous technologies in human detection and counting within predetermined zones necessitated this research. Manual monitoring or simple threshold-based procedures are frequently used in traditional systems, which are time-consuming, error-prone, and lack scalability. YOLOv5 integration tackles these restrictions by providing real-time, accurate, and efficient solutions [15]. The study addresses the increased demand for intelligent systems capable of automating detection and counting processes, allowing for effective resource allocation, risk mitigation, and proactive decision-making in various sectors.

This research bridges the gap between theoretical and practical applications by utilizing advances in deep learning and object detection, providing substantial benefits to companies and organizations functioning in dynamic and complex contexts.

9.1.1 The motivation for the research

The inspiration for this study derives from the growing demand for accurate and real-time person detection and counting inside predetermined zones across multiple domains. Traditional crowd monitoring and management approaches, ensuring security and improving public safety, sometimes fail to deliver effective and dependable solutions. Introducing the deep learning-based object detection framework YOLOv5 provides an opportunity to remove these constraints and revolutionize the sector.

The fundamental aim is to overcome the difficulties associated with traditional approaches, which are frequently labor-intensive, time-consuming, and error-prone. Manual monitoring and counting people within predetermined zones take much time and can be subjective. Furthermore, in dynamic contexts where rapid decision-making is crucial for effective crowd management, security monitoring, and public safety, the necessity for real-time analysis has become critical.

Another driver is the increased demand for proactive measures and enhanced situational awareness across multiple domains. Organizations and authorities can identify possible security risks, reduce congestion, optimize resource allocation, and respond quickly to emergencies by reliably detecting and counting people inside established zones. Furthermore, the potential for technological breakthroughs to streamline operations and improve efficiency in several areas serves as inspiration.

Motivation is also provided by the need to bridge the gap between theoretical advances in deep learning and their practical application. While deep learning-based object identification models have demonstrated exceptional performance on generic object detection tasks, adapting them to specific domains and preset zones necessitates concerted research and development.

9.1.2 Objectives of the research

- The study aims to develop a tool that provides real-time insights to decision-makers, allowing them to take preventative actions and make educated judgments based on trustworthy data.
- The work aims to develop a robust and scalable solution for person detection and counting by exploiting YOLOv5, known for its speed

and accuracy. This technology can potentially improve crowd control at public events, optimize city traffic flow, improve retail analytics, and reinforce security measures in key infrastructure.

- This study tries to apply deep learning-based object detection models to handle human recognition and counting inside predetermined zones, resulting in a realistic and successful solution for real-world applications.

9.1.3 Implications of the research

The outcomes of this study on person detection and counting within predetermined zones using YOLOv5 have wide-ranging consequences. The research serves critical needs in the following categories by correctly monitoring and counting people inside specific areas:

Crowd control: by keeping track of and counting individuals in particular areas, crowd control improves crowd management at public gatherings, stadiums, and transit hubs, helping to ease congestion and quickly handle situations.

Security surveillance: enhances security procedures and response times by enabling security services to monitor for unauthorized access and suspicious activity in restricted areas more effectively.

Public safety: encourages adherence to safety regulations in urban areas and public locations, facilitating quick identification of safety lapses, emergency response, and preventative measures.

Retail analytics: offers insightful information for optimizing store layouts, enhancing consumer satisfaction, and boosting profitability in the retail sector.

Traffic management: improves traffic management in urban areas by analyzing pedestrian movement, congestion hotspots, and personal identification in defined zones, as well as traffic signal timings and pedestrian safety.

9.2 RELATED WORK

The challenges in feature extraction for crowd estimation in exposition centers are analyzed to address the issues of crowd distribution. A crowd density estimation approach has been presented by applying deep learning for passenger flow monitoring systems [16]. To get more information about crowd groups, the pixel difference symbol feature is first extracted, along with the difference amplitude feature and the gray feature of the central pixel, to create the CLBP feature. A dense block model for LR-CNN crowd density estimate was created. According to experimental results, the mean absolute error (MAE) and mean square error (MSE) of the suggested approach in the UCF_CC_50 dataset is finally revealed.

It has been examined how machine learning and artificial intelligence have been used to transform manual, traditional crowd-counting methods into intelligent ones [17]. Convolutional neural networks (CNNs) are highlighted as a promising technique for intelligent picture crowd counting and analysis. Their drawbacks, unique qualities, performance assessment, prospective applications, and future research directions are discussed. Additionally, the article explores the integration of crowd counting with IoT for crowd understanding in smart cities.

A thorough investigation of crowd analysis and management has been conducted, focusing on cutting-edge techniques that can be used in confined and unconstrained settings [18]. The paper discusses the strengths and limitations of these methods, ranging from early seminal research to the latest deep-learning approaches. A comparison of earlier approaches and a thorough discussion of potential future research areas are provided.

There have been several studies on crowd analysis, which can be categorized into two primary branches: crowd statistics and crowd behavior analysis [19]. While crowd behavior analysis looks at the mobility patterns and activities seen in the crowd, crowd statistics evaluate the level of service (LoS) of a crowded environment. The study's objective is to pinpoint underserved or infrequently discussed crowd analysis subfields, particularly when viewed through the perspective of deep learning.

A popular object detection model, YOLOv3, known for its speed and accuracy, has been introduced in [20]. The YOLOv3 architecture forms the basis for subsequent versions, including YOLOv5 and YOLOv8. The research validates the usefulness of the YOLO series in real-time object identification tasks and shows improvements in the series.

Another significant edition of the YOLO series that boasts the best object-detecting speed and accuracy is called YOLOv4 [21]. Although not directly related to YOLOv5 and YOLOv8, the paper [22] provides insights into the continuous advancements in the YOLO family, motivating the exploration of subsequent versions for person detection and counting in predefined zones.

A large mini-batch object detector, MegDet, has been introduced [23]. It addresses the challenge of training object detection models with large mini-batches, which is crucial for achieving efficient person detection and counting in predefined zones. The insights from this work can inform the training strategies and optimization techniques applied to YOLOv5 and YOLOv8 for improved performance.

A single-shot multibox detector (SSD), a popular object detection model, is introduced in this study [24]. Although not specifically focused on YOLOv5 or YOLOv8, SSD provides valuable insights into single-shot detection frameworks and the challenges associated with accurate person detection. The concepts from SSD can complement the research on person detection and counting in predefined zones using YOLO-based models.

The research on Cascade R-CNN, an object detection model that achieves high-quality detection results by employing a cascade of networks, has been presented in [25]. While not directly related to YOLO-based models, the insights from this work can inform strategies for improving the precision and localization accuracy of person detection and counting in predefined zones.

DeepLab, a semantic image segmentation model that uses fully linked conditional random fields (CRFs) and deep convolutional networks, has been introduced in [26]. Although not directly related to person detection, the techniques presented in this work can be valuable for refining the segmentation and localization aspects of person detection and counting within predefined zones using YOLO-based models.

Squeeze YOLO-based people counting (S-YOLO-PC), a new real-time approach for counting individuals, is suggested [27]. The network structure is optimized by S-YOLO-PC using the fire layer of SqueezeNet, which lowers the number of parameters utilized in the model without lowering its precision.

People counting has been employed extensively throughout life, especially on trains, planes, and other forms of public transportation. Service providers manually count the number of customers using a hand counter. This strategy takes a lot of time and effort in the modern world. Therefore, research is suggested so that the service operator can use image processing with the you only look once (YOLO) method instead of counting manually with a hand counter [28].

Understanding the percentage of total traffic that your region draws can be helped by controlling people's behavior in public areas. The YOLOv3 method is utilized to identify people, while background subtraction with Gaussian mixture techniques and the heatmap color technique are used to identify places with a high density of people [29].

9.3 METHODOLOGY

9.3.1 Dataset under study

COCO dataset is used for the experimental analysis. A large-scale dataset for object detection, segmentation, and captioning tasks is called COCO (common objects in context). More than 330,000 photos with annotations for 80 different object categories are included. Deep learning models are frequently trained using it for computer vision problems [30]. In this research, a particular video frame for person identification is considered for the study.

9.3.2 YOLOv5 inference pipeline

The object detection model known as YOLOv5, sometimes called YOLO version 5, has significant improvements over the earlier iterations of YOLO. The inference pipeline of YOLOv5 consists of several steps that enable the

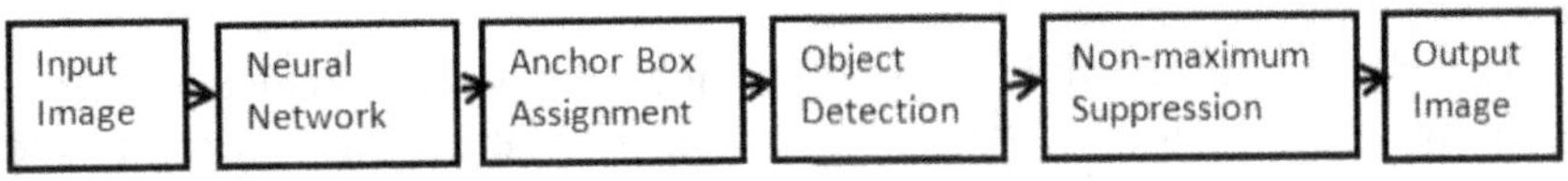

Figure 9.1 YOLOv5 inference pipeline.

model to detect and classify objects in an image. Figure 9.1 provides an overview of the YOLOv5 inference pipeline:

i. **Input image:** the pipeline starts with an input image containing objects that must be detected. The image is frequently preprocessed by scaling it to the required input size for the YOLOv5 model.
ii. **Forward pass of the neural network:** the YOLOv5 neural network, which consists of convolutional layers, down-sampling layers, and detecting layers, is fed to the preprocessed image. The network extracts features from the image at several scales using convolutional procedures.
iii. **Anchor box assignment:** anchor boxes are predefined bounding box shapes with varying aspect ratios and sizes used in YOLOv5. The model assigns anchor boxes to distinct image sections based on the expected features. Each anchor box can detect items of varying sizes and aspect ratios.
iv. **Object detection:** YOLOv5 detects objects in a picture by estimating their bounding boxes and class probabilities. The model generates a series of bounding boxes, each of which is assigned a confidence score indicating the possibility of having an object of interest. Furthermore, class probabilities are projected for each bounding box, showing the likelihood of belonging to various item classes.
v. **Non-maximum suppression (NMS):** YOLOv5 employs a technique known as non-maximum suppression to reduce duplicate and overlapping bounding box predictions. NMS removes bounding boxes based on their confidence scores and the extent to which they overlap, retaining only the most confident and non-overlapping boxes.
vi. **Output:** a set of bounding boxes, each with a class label and a confidence score, are generated using the YOLOv5 inference pipeline. These bounding boxes display the objects identified in the input image and the associated class predictions.

9.3.3 Detectron2 module

Facebook AI Research (FAIR) created the 'detectron2' module, which is a deep learning framework and library for object identification and segmentation applications. It is a sophisticated and adaptable environment for developing computer vision models and doing research in the area of instance-level recognition.

The 'detectron2' module aims to provide a high-performance and user-friendly interface for constructing cutting-edge object identification and segmentation models. It provides a comprehensive collection of training, assessment, inference tools, pre-defined models, and utilities.

The detectron2 module's major features and purposes are as follows:

(a) **Model architecture:** detectron2 has a modular and flexible design that lets users customize and mix different components to develop object detection and segmentation models. It is compatible with backbone designs, feature extractors, and region proposal networks.
(b) **Data loading and augmentation:** detectron2 provides efficient data loading and data augmentation algorithms to handle large-scale datasets. It offers simple APIs for reading and preparing several dataset formats, such as COCO, Pascal VOC, and custom datasets.
(c) **Training and optimization:** the module enables you to train object detection models using optimization techniques, such as weight decay, learning rate schedules, and stochastic gradient descent (SGD) with momentum. Additionally, it enables distributed training across several GPUs and supports mixed-precision training for quicker convergence.
(d) **Inference and evaluation:** detectron2 includes tools for making inferences on new pictures or videos using trained models. It contains post-processing functions for the model's output, such as using non-maximum suppression (NMS) to filter redundant detections. The module also includes metrics for evaluating the performance of object identification and segmentation models.
(e) **Model zoo and pre-trained models:** detectron2 has a model zoo that houses a collection of pre-trained models. These pre-trained models can be applied as a foundation for transfer learning or as a benchmark for evaluation. The module allows you to load these pre-trained models for further customization or inference.

9.3.4 Supervision module

Training data is labeled or annotated by a supervision module to give ground truth information for model training. This module may include either manual labeling by human annotators or automated annotation generation techniques.

The overall process flow of the proposed work is shown in Figure 9.2

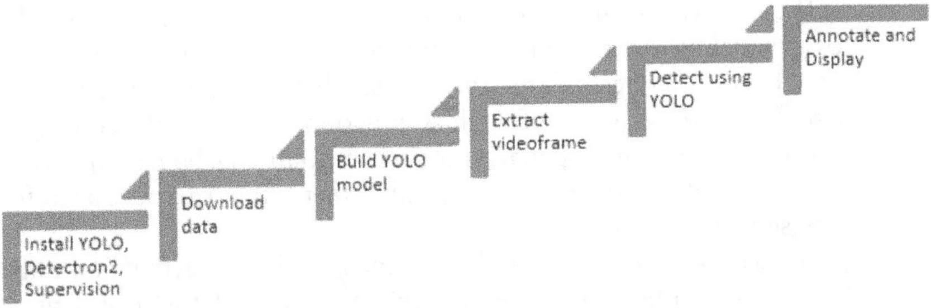

Figure 9.2 Proposed workflow of the implementation.

Python and supporting libraries were used to implement the experiment. Nvidia Cuda Compiler driver with GPU and torch were set up for execution. The YOLO model was trained with the weights of the YOLOv5 large model.

9.4 RESULTS AND DISCUSSION

A video of a marketplace from the COCO dataset is used for the experimental investigation. Initially, a video frame is extracted, and the number of people in the frame is counted using the YOLO model.

9.4.1 Features of YOLOv5

YOLOv5 (you only look once version 5) is effective in detecting objects because of a number of important elements:

- A more intricate architecture known as the EfficientDet network architecture is used by YOLOv5. This architecture increases accuracy and improves generalization to a larger variety of object types.
- A more expansive and varied dataset named D5, which has 600 item types, was used to train YOLOv5.
- YOLOv5 employs a new technique for creating 'dynamic anchor boxes.' The ground truth bounding boxes are first grouped into clusters using a clustering method, and then the centroids of those clusters are used as the anchor boxes. As a result, the anchor boxes can match the size and shape of the identified objects more closely.
- In addition, YOLOv5 also introduces the idea of 'spatial pyramid pooling' (SPP), a pooling layer used to scale down the spatial resolution of the feature maps. SPP enables the model to view the items at various scales, which improves the detection performance for small objects.

- To help the model perform better on datasets with imbalances, YOLOv5 incorporates a new concept termed 'CIoU (Complete IoU) loss,' which is a variation of the IoU loss function. IoU (Intersection over Union) loss calculates the overlap between the expected and actual bounding boxes. This loss function is frequently employed in object detection to penalize incorrect bounding box predictions. However, it does not consider several elements that can be crucial for precise localization.
- By imposing a heavier penalty on incorrectly predicted bounding boxes for smaller items, the area-related term in CIoU loss aids in resolving issues with class imbalance. This is particularly helpful when working with unbalanced datasets that may underrepresent small objects.
- YOLOv5 is renowned for its real-time processing speed, making it far quicker than its forerunners. Through a simplified architecture, it is able to process images quickly, even on simpler hardware.
- In a single pass through the neural network, YOLOv5 completes object detection. This implies that, unlike other object detection methods, it does not require re-analyzing an image at various scales. This lessens duplication and boosts effectiveness.

9.4.2 Comparative study of YOLO5 with state-of-the-art models

A comparative study of the performance of YOLOv5 with other related works is done [12], and the observations are depicted in Table 9.1. The evaluation metrics used for comparing the performance of the algorithms

Table 9.1 Performance of YOLO with other deep learning models

Reference	*Dataset*	*Algorithms*	*Observation*
Li et al., 2021 [22]	Remote sensing images	Faster R-CNN SSD YOLOv3	In comparison to SSD and Faster R-CNN algorithms, YOLO has higher mAP and FPS
Benjdira et al., 2019 [31]	UAV dataset	YOLOv3 Faster R-CNN	YOLOv3 outperforms Faster R-CNN in terms of F1 score and FPS.
Zhao et al., 2019 [32]	Google Earth and DOTA dataset	Faster R-CNN SSD YOLOv3	Compared to Faster R-CNN and SSD, YOLOv3 has higher mAP and FPS.
Ge et al., 2021 [33]	MS COCO dataset	YOLOv3 YOLOv4 YOLOv5	mAP is higher in YOLOv5 than YOLOv3 and YOLOv

Table 9.2 Architecture comparison of YOLO models

	YOLOv3	*YOLOv4*	*YOLOv5*
Neural network	Fully convolutional model	Fully convolutional model	Fully convolutional model
Feature extractor	Darknet-53	CSPDarknet53	CSPDarknet53
Loss function	Binary cross entropy	Binary cross entropy	Binary cross entropy and Logits loss function

are mAP (mean average precision) and FPS (frame per second). The average precision (AP) of all classes in the dataset is considered when computing the mean precision (mAP). FPS is inversely correlated with the time needed to process a single video frame.

The architectures of YOLOv5 with previous versions are compared [12] and tabulated in Table 9.2. YOLOv5's accuracy on PCs and high-end computers is higher than that of YOLOv4 and YOLOv3, with only a slight speed reduction.

Figure 9.3(a) depicts the output recognized by the YOLOv5 model before setting the optimal confidence level. The total number of detections is shown as 110. It is observed that a lot of irrelevant detections are found. All those who do not belong in the person class should be removed. In addition, all detections with a confidence level under 0.5 can be rejected simultaneously. Upon restricting the detection class to a person and setting the confidence level greater than 0.5, the output is depicted in Figure 9.3(b). The total number of person detections was shown as 83. On visual interpretation, five people are not detected by the model. The accuracy of the model for this particular frame was 94.3%.

9.5 PERSON IDENTIFICATION IN RESTRICTED ZONES

Polygons were used to define zones. The coordinates are manually identified, a polygon is constructed, and the number of people within the restricted zone is determined. Figure 9.4 reveals the efficiency of YOLO in human identification in restricted zones 1 and 2 with predefined boundaries. Figure 9.5 reveals the person's identification with complex boundaries.

The accuracy of the model in zonal predictions is tabulated in Table 9.3. The average performance in zonal regions is 86%. It is observed that the objects under minimal lighting conditions are missed in classification. The trained YOLO model performed well, and the confidence threshold was adjusted to improve the accuracy.

(a) (b)

Figure 9.3 Person identification in a market frame (a) before setting optimal confidence level (b) after setting optimal confidence level.

(a) (b)

Figure 9.4 Human identification in the restricted regions: (a) zone 1; (b) zone 2.

(a) (b)

Figure 9.5 Person identification in the complex zones: (a) zone 3 (4); (b) zone 4 (6).

Table 9.3 Performance of the model in restricted zones

Zone	*Correct classification*	*Missed classification*	*Accuracy %*
Zone 1	15	5	75.00
Zone 2	34	7	82.93
Zone 3.1	15	4	78.95
Zone 3.2	4	2	66.67
Zone 3.3	30	0	100.00
Zone 3.4	31	2	93.94
Zone 4.1	9	2	81.82
Zone 4.2	8	2	80.00
Zone 4.3	35	1	97.22
Zone 4.4	4	1	80.00
Zone 4.5	8	0	100.00
Zone 4.6	11	0	100.00

9.6 CONCLUSION

In this study, the features of the YOLOv5 model have been discussed. YOLOv5 is recognized for its real-time processing speed, which makes it far faster than its predecessors. It can process images rapidly on less complex hardware because of a streamlined architecture. YOLOv5 accomplishes object detection in a single pass through the neural network. According to this, it is not necessary to re-examine an image at different scales, unlike some other object detection techniques. Hence, this work investigates how to apply the cutting-edge object detection model, YOLOv5, to identify and count people within predetermined zones. YOLOv5's remarkable performance and efficiency have significantly increased its popularity in the computer vision community. The outcomes of our research showcase the exceptional performance of YOLOv5 in accurately detecting and counting persons within predefined zones. These models demonstrate superior accuracy, capturing persons even in challenging environments with occlusions and varying scales. The accuracy of the model on the entire market frame was 94.3%. The zonal predictions produce an average accuracy of 86%. The real-time capabilities of YOLOv5 enable prompt detection and counting, empowering decision-makers with timely insights for efficient crowd management and enhanced security measures. Crowd control, security surveillance, and public safety applications have undergone a paradigm shift due to the integration of these cutting-edge object detection frameworks. The ability to accurately monitor and count persons in specific areas provides unprecedented situational awareness, enabling proactive responses and efficient resource allocation in dynamic environments.

REFERENCES

1. Rajendra, P., Mina Kumari, Sangeeta Rani, Namrata Dogra, Rahul Boadh, Ajay Kumar, and Mamta Dahiya. "Impact of artificial intelligence on civilization: future perspectives." *Materials Today: Proceedings* 56 (2022): 252–256.
2. Kumar, Amit, Vinod Kumar, Vikas Modgil, and Ajay Kumar. "Stochastic Petri nets modelling for performance assessment of a manufacturing unit." *Materials Today: Proceedings* 56 (2022): 215–219.
3. Kumar, Ajay, Hari Singh, Parveen Kumar, and Bandar AlMangour, eds. *Handbook of Smart Manufacturing: Forecasting the Future of Industry 4.0.* CRC Press (2023).
4. Joshi, Vishwas Deep, Priya Agarwal, and Ajay Kumar. "Fuzzy transportation planning: a goal programming tactic for navigating uncertainty and multi-objective decision making." *International Journal on Interactive Design and Manufacturing (IJIDeM)* 17 (2023): 1–29.
5. Rani, Sangeeta, Khushboo Tripathi, and Ajay Kumar. "Machine learning aided malware detection for secure and smart manufacturing: a comprehensive analysis of the state of the art." *International Journal on Interactive Design and Manufacturing (IJIDeM)* 17 (2023): 1–28.
6. Rani, Sangeeta, Khushboo Tripathi, Yojna Arora, and Ajay Kumar. "Analysis of anomaly detection of Malware using KNN." In *2022 2nd International Conference on Innovative Practices in Technology and Management (ICIPTM)*, vol. 2, pp. 774–779. IEEE (2022).

7. Rani, Sangeeta, Khushboo Tripathi, Yojna Arora, and Ajay Kumar. "A machine learning approach to analyze cloud computing attacks." In *2022 5th International Conference on Contemporary Computing and Informatics (IC3I)*, pp. 22–26. IEEE (2022).
8. Krishnachaithanya, N., Gurdit Singh, Smita Sharma, Rangisetti Dinesh, Sumeet Ramsingh Sihag, Kamna Solanki, Abhishek Agarwal, Mrinalini Rana, and Ujjwal Makkar. "People counting in public spaces using deep learning-based object detection and tracking techniques." In 2023 *International Conference on Computational Intelligence and Sustainable Engineering Solutions (CISES)*, pp. 784–788. IEEE (2023).
9. Zhao, Kun, and Xiaoxi Ren. "Small aircraft detection in remote sensing images based on YOLOv3." In *IOP Conference Series: Materials Science and Engineering*, vol. 533, no. 1, p. 012056. IOP Publishing (2019).
10. Nugroho, Anan, Faizal Indaryanto, and Alfa Faridh Suni. "Distance and people counting app based on YOLO as a Covid-19 health protocol." In *AIP Conference Proceedings*, vol. 2727, no. 1. AIP Publishing (2023).
11. Ren, Peiming, Wei Fang, and Soufiene Djahel. "A novel YOLO-based real-time people counting approach." In *2017 International Smart Cities Conference* (ISC2), pp. 1–2. IEEE (2017).
12. Nagolu, Hyndhavi, and Santhoshini Sahu. "Empirical study on real time people counting using deep learning." In *2023 10th International Conference on Computing for Sustainable Global Development (INDIACom)*, pp. 1–4. IEEE (2023).
13. Li, Min, Zhijie Zhang, Liping Lei, Xiaofan Wang, and Xudong Guo. "Agricultural greenhouses detection in high-resolution satellite images based on convolutional neural networks: Comparison of faster R-CNN, YOLO v3 and SSD." *Sensors* 20, no. 17 (2020): 4938.
14. Maktoof, Mohammed Abdul Jaleel, Ibraheem Nadher Ibraheem, and Israa Tahseen Ali Al_attar. "Crowd counting using Yolov5 and KCF." *Periodicals of Engineering and Natural Sciences* 11, no. 2 (2023): 92–101.
15. Benjdira, Bilel, Taha Khursheed, Anis Koubaa, Adel Ammar, and Kais Ouni. "Car detection using unmanned aerial vehicles: comparison between faster R-CNN and yolov3." In *2019 1st International Conference on Unmanned Vehicle Systems-Oman (UVS)*, pp. 1–6. IEEE. (2019).
16. Bhangale, Ujwala, Suchitra Patil, Vaibhav Vishwanath, Parth Thakker, Amey Bansode, and Devesh Navandhar. "Near real-time crowd counting using deep learning approach." *Procedia Computer Science* 171 (2020): 770–779.
17. Ilyas, Naveed, Ahsan Shahzad, and Kiseon Kim. "Convolutional-neural network-based image crowd counting: review, categorization, analysis, and performance evaluation." *Sensors* 20, no. 1 (2019): 43.
18. Khan, Khalil, Waleed Albattah, Rehan Ullah Khan, Ali Mustafa Qamar, and Durre Nayab. "Advances and trends in real time visual crowd analysis." *Sensors* 20, no. 18 (2020): 5073.
19. Bendali-Braham, Mounir, Jonathan Weber, Germain Forestier, Lhassane Idoumghar, and Pierre-Alain Muller. "Recent trends in crowd analysis: a review." *Machine Learning with Applications* 4 (2021): 100023.
20. Redmon, Joseph, and Ali Farhadi. "Yolov3: An incremental improvement." arXiv preprint arXiv:1804.02767 (2018).

21. Bochkovskiy, Alexey, Chien-Yao Wang, and Hong-Yuan Mark Liao. "Yolov4: Optimal speed and accuracy of object detection." arXiv preprint arXiv:2004.10934 (2020).
22. Nepal, Upesh, and Hossein Eslamiat. "Comparing YOLOv3, YOLOv4 and YOLOv5 for autonomous landing spot detection in faulty UAVs." *Sensors* 22, no. 2 (2022): 464.
23. Peng, Chao, Tete Xiao, Zeming Li, Yuning Jiang, Xiangyu Zhang, Kai Jia, Gang Yu, and Jian Sun. "Megdet: a large mini-batch object detector." In *Proceedings of the IEEE conference on Computer Vision and Pattern Recognition*, pp. 6181–6189, IEEE Xplore (2018).
24. Liu, Wei, Dragomir Anguelov, Dumitru Erhan, Christian Szegedy, Scott Reed, Cheng-Yang Fu, and Alexander C. Berg. "SSD: single shot multibox detector." In *Computer Vision–ECCV 2016: 14th European Conference, Amsterdam, The Netherlands, October 11–14, 2016, Proceedings, Part I 14*, pp. 21–37. Springer International Publishing (2016).
25. Cai, Zhaowei, and Nuno Vasconcelos. "Cascade r-cnn: delving into high quality object detection." In *Proceedings of the IEEE Conference on Computer Vision and Pattern Recognition*, pp. 6154–6162, IEEE Xplore (2018).
26. Chen, Liang-Chieh, George Papandreou, Iasonas Kokkinos, Kevin Murphy, and Alan L. Yuille. "Deeplab: semantic image segmentation with deep convolutional nets, atrous convolution, and fully connected crfs." *IEEE Transactions on Pattern Analysis and Machine Intelligence* 40, no. 4 (2017): 834–848.
27. Ren, Peiming, Lin Wang, Wei Fang, Shulin Song, and Soufiene Djahel. "A novel squeeze YOLO-based real-time people counting approach." International Journal of Bio-Inspired Computation 16, no. 2 (2020): 94–101.
28. Kusuma, Tsabita Al Asshifa Hadi, Koredianto Usman, and Sofia Saidah. "People counting for public transportations using you only look once method." *Jurnal Teknik Informatika (Jutif)* 2, no. 1 (2021): 57–66.
29. Kajabad, Ebrahim Najafi, and Sergey V. Ivanov. "People detection and finding attractive areas by the use of movement detection analysis and deep learning approach." *Procedia Computer Science* 156 (2019): 327–337.
30. Dwyer, B., J. Nelson, and J. Solawetz. "Roboflow (version 1.0)[software]." (2022). https://roboflow.com/research
31. Ge, Zheng, Songtao Liu, Feng Wang, Zeming Li, and Jian Sun. "Yolox: Exceeding yolo series in 2021." arXiv preprint arXiv:2107.08430 (2021).
32. Gündüz, Mehmet Şirin, and Gültekin Işık. "A new YOLO-based method for real-time crowd detection from video and performance analysis of YOLO models." *Journal of Real-Time Image Processing* 20, no. 1 (2023): 5.
33. Xiang, Jun, and Na Liu. "Crowd density estimation method using deep learning for passenger flow detection system in exhibition center." *Scientific Programming* 2022 (2022): 1–9.

Chapter 10

Revolutionizing cloud computing security with the power of artificial intelligence

Advancements, challenges, and opportunities

Tarun Kumar Vashishth, Vikas Sharma, Kewal Krishan Sharma, Bhupendra Kumar, Sachin Chaudhary, and Rajneesh Panwar

School of Computer Science and Applications, IIMT University, Meerut, India

10.1 INTRODUCTION

CC has significantly transformed the way businesses store and process their data, allowing them to operate more efficiently and cost-effectively. However, the dynamic and distributed nature of CC has posed several security challenges, including data breaches, unauthorized access, and service disruptions. Old ways of security solutions for example as firewalls, anti-virus s/w, and intrusion detection systems (IDS) do not seem enough to protect cloud infrastructures against these threats. Hence, researchers and industry practitioners have turned to AI to enhance cloud security.

AI-based cloud security

AI has the potential to revolutionize cloud security by providing intelligent and automated security solutions. One of the significant benefits of AI-based cloud security solutions is anomaly detection. Anomaly detection involves identifying unusual patterns or activities that can highlight a security breach. AI-based anomaly detection algorithm may identify patterns and behaviors that traditional security solutions may not detect. AI can analyze network traffic and identify not usual patterns that may identify a distributed denial of service (DDoS) attack. Another way in which AI is being used to enhance cloud security is TI. This involves collecting and analyzing information to note potential security threats. AI can be used to analyze vast amounts of data from various sources, including network logs, social media, and dark web forums, to identify emerging threats. For example, AI-based TI algorithms can analyze social media data to identify potential insider threats. Automated incident response (AIR) is another area where AI is being utilized to enhance cloud security. AIR involves using AI-based algorithms to automatically respond to security incidents. Suppose

DOI: 10.1201/9781003502470-10

an AI-based algorithm detects a DDoS threat, it automatically stops traffic from attacking IP address or reroute traffic to a different server to mitigate the attack. Anomaly detection can be achieved using different AI techniques such as machine learning, deep learning, and neural networks. These techniques allow the AI algorithms to know by available historical data and identify signature that may indicate a security breach. ML algorithms can be used to analyze network traffic and identify unusual patterns that may indicate a DDoS threat. DL algorithms may be used to analyze log files and identify unusual patterns that may indicate a potential security threat. TI is another area where AI is being used to enhance cloud security. Threats with intelligence involve collecting and examining information to identify potential security threats. AI can be used to analyze huge data from various sources, including network logs, social web sites, and dark web sites, to pin point emerging threats. For example, AI-based TI algorithms can analyze

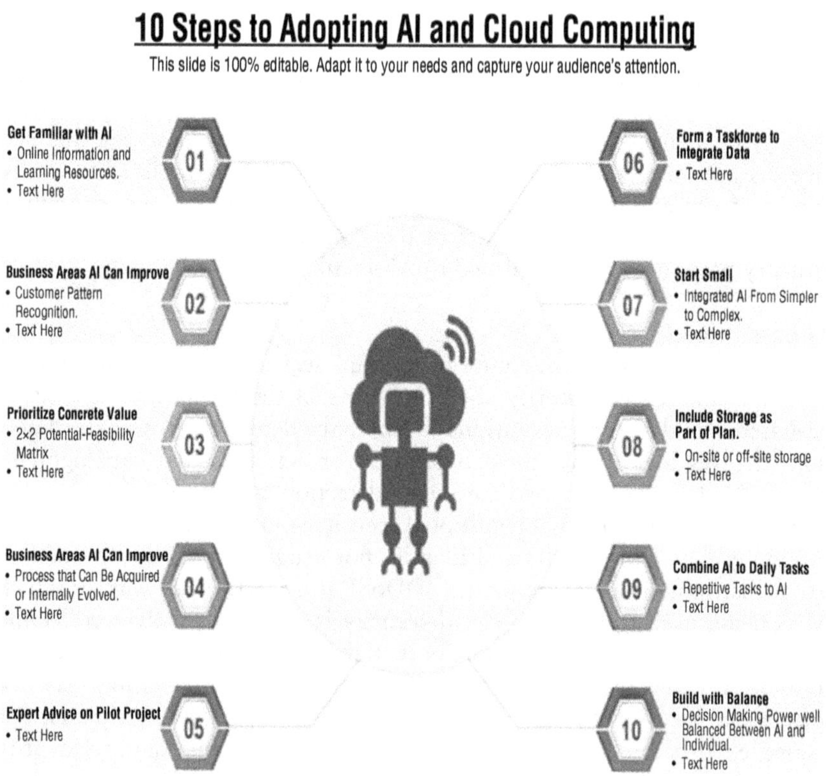

Figure 10.1 Ten steps to adopting AI and cloud computing.

social media data to identify potential insider threats. This is shown in the Figure 10.1

Revolutionizing CC security with the power of artificial intelligence (AI) offers several advantages over traditional security solutions. Some of the key advantages include the follwing:

a. **Enhanced threat detection:** AI-based algorithms can detect threats that traditional security solutions may not identify. AI can analyze vast amounts of data, network traffic, logs, and user behavior, to segregate patterns that can inform a security threat. It helps systems to identify and act to threats quickly, lower risk of a security breach.
b. **Automated incident response:** AI-based algorithms can automate incident reactions, lowering the requirements for human involvements. Automated incident response can help to identify security events quickly and correctness, mitigating impact of the incident.
c. **Reduced human error:** AI-based algorithms can reduce the risk of human error in security operations. Traditional security solutions often require manual analysis and decision-making, which can be prone to error. AI can automate these processes, reducing the risk of error and improving the accuracy of security operations.
d. **Improved scalability:** AI-based cloud security solutions can be scaled easily, allowing organizations to protect their cloud infrastructures as they grow. Traditional security solutions may struggle to scale as the volume of data and traffic increases, leading to potential security gaps.
e. **Cost-effective:** AI-based cloud security solutions can be more cost-effective than traditional security solutions. AI can automate security operations, reducing the need for manual analysis and intervention. This can lead to cost savings for organizations, particularly those with large and complex cloud infrastructures.
f. **Real-time TI:** AI-based cloud security measures can give on real-time basis TI, allowing organizations to react instantly to emerging attacks. Real-time TI can help organizations to prevent security breaches before they occur, reducing the impact on the organization.

Overall, the advantages of revolutionizing CC security with the power of AI are significant. AI-based solutions can improve threat detection, automate incident response, reduce human error, improve scalability, and provide cost-effective and real-time TI. These benefits make AI-based cloud security solutions an attractive option for organizations looking to enhance their cloud security posture.

Revolutionizing CC security with the power of artificial intelligence (AI) comes with its fair share of challenges. Some of these challenges include the following:

a. **Data privacy:** artificial intelligence relies heavily on large amounts of data to train and enhance its algorithms. However, this data often contains sensitive information, and protecting it from unauthorized access is a major challenge. Cloud providers must ensure that customer data is protected at all times, and AI systems must be model with privacy.
b. **Fairness and bias:** AI is only as upright as the data they are trained on. If the info is altered, then the AI system will be not true as well. It may lead to not fair decisions and discrimination, which may have serious consequences. Cloud providers must ensure, AI systems are designed to be fair enough and not biased.
c. **Complexity:** CC environments are becoming increasingly complex, with many different systems and technologies interacting with each other. AI systems must be able to operate in this complex environment and integrate with other systems seamlessly.
d. **Cybersecurity threats:** cloud providers are often targeted by cybercriminals looking to steal sensitive data. AI systems can be used to detect and prevent these threats, but they must be constantly updated to stay ahead of evolving threats.
e. **Lack of trust:** many organizations are still hesitant to use AI systems for security because they do not fully understand how they work. Cloud providers must be transparent about how their AI systems operate and provide clear explanations for any decisions made by the system.
f. **Resource constraints:** AI systems require significant computing resources to operate effectively. Cloud providers must ensure that they have the necessary resources to support their AI systems without impacting the performance of other systems.

Overall, revolutionizing CC security with AI is a complex undertaking that requires careful planning and execution. Cloud providers must be known of these challenges and take corrective measurements to mitigate them to ensure the security and privacy of such customers' data.

10.2 LITERATURE REVIEW

The article titled 'Revolutionizing CC Security with the Power of Artificial Intelligence: Advancements, Challenges, and Opportunities' explores the involvement of AI in enhancing the safety of CC systems. FThe following is a literature review of the article.

The article begins with an overview of the challenges faced by CC security and the potential for AI to address these challenges. The authors note that while CC offers numerous benefits, it also introduces new security risks, including data breaches, insider threats, and denial of service (DoS) attacks. AI, the authors argue, can help to mitigate these risks by analyzing huge amount of data in real-time with identifying critical security threats before they take place. The authors then discuss several advancements in AI that are relevant to CC security. These include machine learning algorithms, which can learn to identify patterns and anomalies in large datasets; natural language processing (NLP), which can help to detect and prevent phishing attacks; and deep learning techniques, which can improve the accuracy of threat detection models.

The article also points out some of the issues associated with implementing AI in CC security. One of primary difficulties is the absence of skilful personnel who can implement and develop AI mechanisms. The authors also note that AI models can be vulnerable to attacks themselves, and that it is important to ensure the security of the AI infrastructure.

In [1], Nabila et al. (2021) provides a comprehensive overview of the potential impact of AI and robots on society, highlighting both the potential benefits and challenges. The authors suggest that careful consideration and planning will be necessary to ensure that advantages of AI and robots are maximized while minimizing critical adverse consequences. In [2], Mustapha et al. (2021) provide a comprehensive overview of the potential of CIA technologies in promoting sustainable aquaculture development. The authors suggest that further research and development in such areas will be necessary to fully recognize the strength of such technologies in the aquaculture industry. In [3], Zhang et al. (2021) provide a comprehensive overview of the potential of including blockchain and AI technologies, as well as the challenges and research issues associated with this integration. The authors suggest that research must carry further and development in this area will be necessary to fully realize the criticality of the technologies in different industries. In [4], Kakhi et al. (2022) gave the comprehensive overview of the strength of integrating IoMT and AI technologies for healthcare, as well as the challenges and opportunities associated with this integration. The authors suggest that further research and development in this area will be necessary to fully realize the potential of these technologies in transforming healthcare delivery. In [5], Ahamad et al. (2022) provide a full analysis of the potential of AI in enabling high-performance CC systems, as well as the challenges and research issues associated with this integration. The authors suggest that further research and development in this area will be necessary to all realize the potential of these technologies in various industries. In [6], Ghani et al. (2020) provide several examples of applications of cloud storage systems, such as cloud-based backup and disaster recovery, content distribution networks, and big data analytics. The authors also discuss the potential benefits of these applications, including improved data availability, reduced costs, and increased efficiency. Overall, the article provides a comprehensive overview of the architecture of cloud storage systems, as well as the difficulties and issues associated with their implementation. The authors suggest that research and development in this area will be necessary to further fully realize the potential of cloud storage systems. In [7], Ilager et al. (2020) discuss the critical advantage of using AI in resource management, for example as enhanced efficiency, increase performance, and reduced costs. They provide several examples of AI-based resource management techniques, such as reinforcement learning and deep learning, and discuss their potential applications in various industries. In [8], Schmidt et al. (2021) provide specific recommendations for the use of AI in various areas of national security, such as intelligence, defense, and cybersecurity. For example, it recommends the development of AI tools to improve intelligence gathering and analysis, as well as the use of AI for autonomous systems and unmanned vehicles. Overall, the report provides a detailed analysis of the opportunities and challenges associated with AI for national security, as well as a roadmap for the US to leverage this technology in a strategic and effective manner.

In [9], Prabavathy et al. (2022) highlight several challenges related to the adoption of smart technologies in healthcare, such as the need for interoperability standards and data privacy and security concerns. They argue that addressing these challenges is critical to realizing the full potential of smart technologies in healthcare. In [10], Matthew et al. (2021) give a full detail of h/w and s/w components of the system, including UAV platform, the sensors used for data acquisition, and the AI algorithms used for data analysis and decision-making. The authors also present the results of experiments conducted to evaluate the performance of the system in real-world surveillance scenarios. Overall, the article demonstrates the potential of using autonomous UAVs with AI capabilities for security surveillance applications. The authors argue that such systems can provide a cost-effective and efficient means of gathering and analyzing data for security purposes, and they suggest that further research in this area could lead to significant advances in security surveillance technology. In [11], Hernandez et al. (2023) give an important resource for researchers and practitioners interested in the intersection of CC and intelligent vehicles. The authors suggest that future research in this area should focus on developing new cloud-based architectures and algorithms that can address the unique challenges of intelligent vehicle applications, and that can help to accelerate the adoption of these technologies in real-world settings. In [12], Benlian et al. (2018) argue that CC enables organizations to decouple their IT resources from physical infrastructure, leading to increased flexibility and scalability. The platformization process involves creating and leveraging common platforms that enable the development and deployment of new services and applications. Finally, the recombination process involves combining various platforms to create innovative solutions. The authors provide examples of companies that have successfully leveraged CC to create transformative value and discuss the implications for future research and practice. In [13], Rajendra, et al. (2022) delve into the profound influence of artificial intelligence on society and its potential ramifications on civilization. The authors explore the far-reaching implications of AI in various domains, contemplating its implications on socio-cultural, economic, and technological aspects. They provide future-oriented perspectives, anticipating the transformative role AI may play in shaping our societies. In [14], Rani et al. (2021) focus on the development and implementation of RPL (routing protocol for low-power and lossy networks) based routing protocols. Specifically, it addresses the challenge of load balancing in IoT (Internet of Things) networks. The authors present a novel approach that contributes to the efficiency and stability of IoT network communications. In [15], Jain and Sharma (2022) introduce a quantum-inspired hybrid algorithm, combining simulated annealing (SSA) and grey wolf optimizer (GWO), tailored for service level agreement (SLA) based task scheduling in CC environments. The proposed approach aims to enhance quality of service (QoS) parameters, thereby improving the overall efficiency and performance of cloud services. In [16], Phogat et al. (2021) presents an innovative approach to automating irrigation techniques utilizing artificial neural networks and fuzzy logic. The authors address the critical need for precise and efficient irrigation methods in agriculture. The combination of these advanced computational techniques shows promise in optimizing water resource management for agricultural purposes. In [17], Keshav and Sangeeta (2023) focus on the application of the K-nearest neighbor (KNN) algorithm for the analysis of anomaly detection in malware. The authors investigate the effectiveness of this approach in identifying and mitigating malicious software. The findings contribute to the field of cybersecurity, addressing the pressing need for robust malware detection mechanisms. In [18], Rani et al. (2022) presents a machine learning-based approach for analyzing and mitigating attacks on CC infrastructures. The authors employ advanced techniques to identify and respond to security threats in cloud environments. The research is significant in safeguarding critical digital assets and ensuring the integrity of cloud services. In [19], Jain et al. (2022) introduces a novel optimization algorithm inspired by the behavior of salp swarms. The algorithm is tailored for resource optimization in CC environments, with specific consideration for constraints related to time and cost. The proposed approach aims to enhance the efficiency and cost-effectiveness of resource allocation in cloud infrastructures.

Finally, the authors discuss several opportunities for next research and development in the area. These include improving the accuracy and efficiency of AI models for threat detection, developing more advanced AI techniques for analyzing complex data, and exploring the use of AI in other areas of CC security, such as access control and identity management. Table 10.1 shows the overview of the key details and focuses of each research paper considered in the literature.

Overall, the article provides a comprehensive review of the potential for AI to revolutionize CC security. While there are certainly challenges to be addressed, the authors demonstrate that AI consists of the potential to significantly improve the security of CC systems, and that there are numerous opportunities for the next research and development in the area.

10.3 DISCUSSION

One proposed methodology to improve cloud security using artificial intelligence (AI) involves the following steps:

a. **Data gathering:** the first step is to gather data from different sources, including logs – network, logs – system, user behavior info, and external TI feeds.
b. **Data pre-processing:** next, the collected data needs to be pre-processed to convert it into a format that can be ingested by AI algorithms. This may include works like data cleaning, normalization, and extraction of feature.
c. **AI model training:** once data is already processed, AI models may be skilled on the data to point out patterns and anomalies that may indicate security threats. It may involve using supervised, unsupervised, or semi-supervised learning techniques, depending on the nature of the data and the specific security threats being addressed.
d. **Real-time monitoring:** once the AI models are trained, they can be deployed to monitor cloud environments in real-time. This involves ingesting live data from various sources and using the trained models to identify potential threats. The AI models can also be used to generate alerts or trigger automated responses when potential threats are detected.
e. **Continuous improvement:** finally, the AI models should be continuously updated and improved based on feedback from the real-time monitoring system. This may involve retraining the models on new data, fine-tuning parameters, or adding new features to improve accuracy and reduce false positives. This is shown in the Figure 10.2

By following this methodology, cloud providers can leverage the power of AI to identify and act to security attacks in a real-time manner, improving overall security posture to their cloud environments.

Table 10.1 This table provides a concise overview of the key details and focuses of each research paper

Paper title and authors	*Publication year*	*Venue/journal*	*Key focus*
Impact of AI on Civilization: Future Perspectives by Rajendra, P., Kumari, M., Rani, S., Dogra, N., Boadh, R., Kumar, A., & Dahiya, M.	2022	Materials Today: Proceedings, 56, 252–256	Influence of AI on society, future perspectives.
RPL based routing protocols for load balancing in IoT network by Rani, S., Kumar, A., Bagchi, A., Yadav, S., & Kumar, S.	2021	Journal of Physics: Conference Series, Vol. 1950, No. 1, p. 012073 (IOP Publishing)	IoT network load balancing with RPL-based protocols.
Quantum inspired hybrid SSA–GWO algorithm for SLA based task scheduling by Jain, R., Sharma, N.	2022	Cluster Computing	SLA-based task scheduling in CC with SSA-GWO algorithm.
Automating irrigation techniques based on ANN and fuzzy logic by Phogat, M., Kumar, A., Nandal, D., & Shokhanda, J.	2021	Journal of Physics: Conference Series, Vol. 1950, No. 1, p. 012088 (IOP Publishing)	Irrigation automation using ANN and fuzzy logic.
Analysis of Anomaly detection of Malware using KNN by Rani, S., Tripathi, K., Arora, Y., & Kumar, A.	2022	2022 2nd International Conference on Innovative Practices in Technology and Management (ICIPTM), Vol. 2, pp. 774–779 (IEEE)	Malware anomaly detection using K-Nearest Neighbours.
Machine Learning Approach to Analyze CC Attacks by Rani, S., Tripathi, K., Arora, Y., & Kumar, A.	2022	2022 5th International Conference on Contemporary Computing and Informatics (IC3I), pp. 22–26 (IEEE)	Machine learning for analysingCC attacks.
Salp Swarm Algorithm for Resource Optimization in CC by Jain, R., & Sharma, N.	2022	International Journal of Applied Metaheuristic Computing (IJAMC), 13(1), 1–21	Salp swarm algorithm for resource optimization in CC.

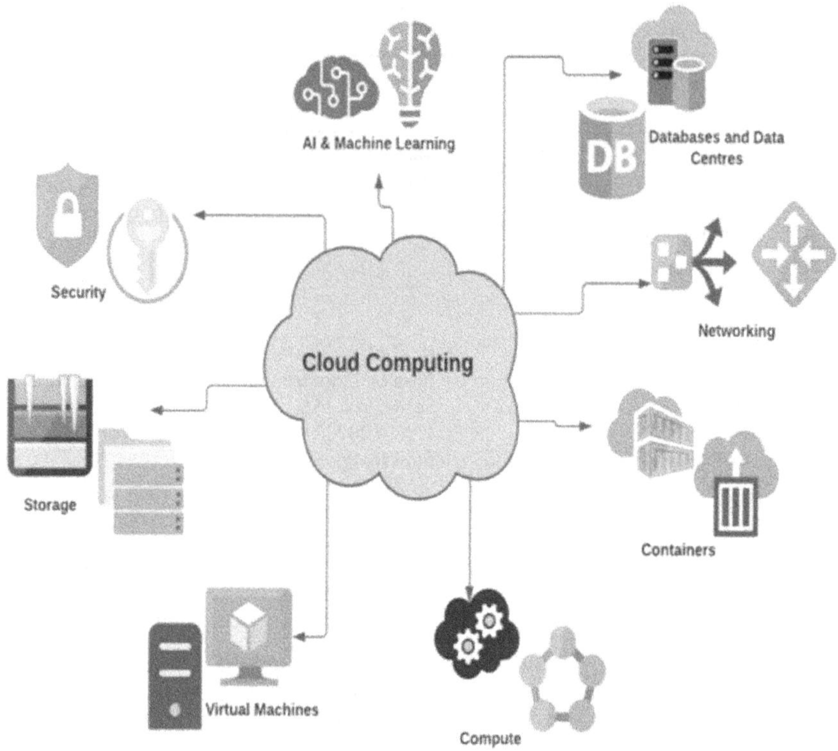

Figure 10.2 The cloud computing sector is divided into key related sectors.

10.4 TABLE/FORMULATION

The formula for enhancing cloud security using AI can be represented as

$$S = f(D, M, R)$$

Using the sample values (see Table 10.2), it may calculate the result of the formula as follows:

$S = f(D, M, R)$
$S = k * (D * M) * R$ (where k is a constant)
S = 1 * (High * Medium)* High (assuming equal weightage to all variables)
S = High *(Medium * High)
S = High

Table 10.2 Various variable and related sample values

Symbol	*Description*	*Sample value*
S	The level of cloud security	
D	The data being processed and stored in the cloud	High
M	The AI models and algorithms used for security	Medium
R	The resources available for implementing security	High
f()	The function that represents the relationship between variables	

Therefore, the level of cloud security (*S*) is high for these sample values, indicating that the combination of high data sensitivity, medium AI models and algorithms, and high resources for implementing security can result in a high level of cloud security.

Where *S* represents the overall security level of the cloud environment, *D* represents the data collected from various sources, *M* represents the AI models used to analyze the data and detect potential threats, and *R* represents the response mechanisms that are triggered when threats are detected. The function *f* represents the complex set of algorithms and processes used to transform the raw data into actionable insights and responses. This may involve tasks such as data pre-processing, feature selection, model training, and real-time monitoring.

To enhance cloud security using AI, the following steps can be taken:

- Collect info from various sources, including logs of network, system logs, user behavior data, and external TI feeds.
- Pre-process the data to convert it into a format that can be ingested by AI algorithms. It may involve works for example as data-cleaning, normalization, and feature extraction.
- Train AI models on the pre-processed information to know patterns and anomalies that can segregate security threats. It may involve using supervised, unsupervised, or semi-supervised learning techniques, depending on the nature of the data and the specific security threats being addressed.
- Deploy the AI models to monitor cloud environments in real-time. This involves ingesting live data from various sources and using the trained models to identify potential threats.
- Trigger automated responses when potential threats are detected. This may involve tasks such as blocking traffic, quarantining infected machines, or alerting security teams.

By following this formula, cloud providers can leverage the power of AI to enhance the security of their cloud environments and protect against a wide variety of attacks, including malware, adware, distributed DoS attacks, and threats from inside.

10.5 CONCLUSION

The integration of AI in CC security has revolutionized the way organizations protect their data and infrastructure from potential threats. AI-powered cloud security solutions have become increasingly popular because their ability to provide real-time threat detection and response, improve security posture, and lower risk of security breaches. However, inclusion of AI in cloud security also poses some challenges such as the need for skilled personnel, ethical concerns, and the risk of over-reliance on technology. To fully leverage the power of AI in cloud security, organizations need to address these challenges and adopt a holistic approach that combines human expertise, AI technology, and appropriate security protocols. The advancements in AI and cloud security offer significant opportunities for organizations to improve their security posture, enhance their competitive advantage, and provide better customer experiences. As the reliance on CC continues to grow, the role of AI in cloud security is expected to become even more critical, and organizations need to stay informed and adapt to the evolving threat landscape to stay protected.

Integration of AI into CC security marks a significant milestone in fortifying the digital infrastructure of organizations worldwide. This synergy capitalizes on AI's prowess in data analysis, pattern recognition, and adaptive learning to proactively identify and mitigate emerging threats. The advancements made in this field have ushered in a new era of cyber-resilience, empowering businesses to navigate the intricate landscape of cloud security with greater confidence.

The deployment of AI-driven anomaly detection systems and predictive analytics has exhibited remarkable success in pre-emptively identifying suspicious activities, thereby averting potential breaches. Furthermore, the adaptive nature of AI algorithms ensures continuous refinement, enabling them to evolve alongside evolving threat vectors. This dynamic approach is a formidable defence against the ever-evolving tactics employed by cyber adversaries.

However, while the promise of AI in cloud security is immense, it is not without its challenges. Ethical considerations surrounding data privacy, algorithm bias, and transparency remain paramount. Striking a balance between innovations and safeguarding individual rights will be pivotal in shaping the ethical framework of AI-powered security solutions.

Despite these challenges, the opportunities presented by the convergence of AI and cloud security are monumental. The potential for proactive threat hunting, rapid incident response and adaptive defence strategies promises to reshape the cybersecurity landscape. Moreover, the democratization of AI technologies through cloud services democratizes access to state-of-the-art security capabilities, empowering a wider range of organizations to bolster their defences.

In conclusion, overall, the marriage of AI and CC security represents a paradigm shift in safeguarding digital assets. The ongoing research, development, and collaboration between industry, academia, and policymakers will be instrumental in harnessing the full potential of this transformative union.

10.6 FUTURE SCOPE

While AI keeps on going to improve and be effective, it can be assumed to see even more innovative solutions for cloud security. Some potential future developments in this field could include the following:

- **Explainable AI:** while AI trends are more difficult, it becomes complex to understand how it can make it decisive. Explainable AI aims to tackle how AI systems work and arrive at their conclusions. This could help build trust in AI-powered security systems and make them more transparent and understandable.
- **Deep learning:** DL is sub-set of AI that involves training neural networks to learn from data. Deep learning algorithms can analyze vast amounts of data to identify patterns and make predictions. It could be used in cloud security to find and react to attacks more quickly and accurately.
- **Natural language processing (NLP):** NLP is a branch of AI that involves analyzing language of humans. NLP algorithms may be used to analyze security logs and identify potential threats based on the language used. This could help organizations find and react to security incidents more quickly.
- **Edge computing:** edge computing consist processing information-data near to origination of generation, rather than sending it to the cloud. It could be used in cloud security to detect threats at the edge and respond in real-time, reducing the time it takes to find and react to security incidents.
- **Quantum computing:** this type of computing is an up-coming technology that could revolutionize computing by processing vast amounts of data more quickly than traditional computers. This could be used in cloud security to quickly analyze large amounts of data and detect potential threats more quickly.

Overall, the future of cloud security with AI looks bright. Since AI keeps on developing and maturing, we can expect to see new innovation outcomes to protect cloud environments from emerging threats.

REFERENCES

1. Nabila, Efa Ayu, Sugeng Santoso, Yudi Muhtadi, and Budi Tjahjono. 'Artificial Intelligence Robots and Revolutionizing Society in Terms of Technology, Innovation, Work and Power'. *IAIC Transactions on Sustainable Digital Innovation (ITSDI)* 3, no. 1 (5 November 2021): 46–52. https://doi.org/10.34306/itsdi.v3i1.526.
2. Mustapha, Umar Farouk, Abdul-Wadud Alhassan, Dong-Neng Jiang, and Guang-Li Li. 'Sustainable Aquaculture Development: A Review on the Roles of Cloud Computing, Internet of Things and Artificial Intelligence (CIA)'. *Reviews in Aquaculture* 13, no. 4 (September 2021): 2076–91. https://doi.org/10.1111/raq.12559.
3. Zhang, Zhonghua, Xifei Song, Lei Liu, Jie Yin, Yu Wang, and Dapeng Lan. 'Recent Advances in Blockchain and Artificial Intelligence Integration: Feasibility Analysis, Research Issues, Applications, Challenges, and Future Work'. *Security and Communication Networks* 2021 (24 June 2021): 1–15. https://doi.org/10.1155/2021/9991535.
4. Kakhi, Kourosh, Roohallah Alizadehsani, H. M. Dipu Kabir, Abbas Khosravi, Saeid Nahavandi, and U. Rajendra Acharya. 'The Internet of Medical Things and Artificial Intelligence: Trends, Challenges, and Opportunities'. *Biocybernetics and Biomedical Engineering* 42, no. 3 (July 2022): 749–71. https://doi.org/10.1016/j.bbe.2022.05.008.
5. Ahamad, Shahanawaj, Milad Mohseni, Vineet Shekher, Ghassan Fadhil Smaisim, Amrendra Tripathi, and Joel Alanya-Beltran. 'A Detailed Analysis of the Critical Role of Artificial Intelligence in Enabling High-Performance Cloud Computing Systems'. In *2022 2nd International Conference on Advance Computing and Innovative Technologies in Engineering (ICACITE)*. IEEE (2022). https://doi.org/10.1109/icacite53722.2022.9823679.
6. Ghani, Anwar, Afzal Badshah, Saeedullah Jan, Abdulrahman A. Alshdadi, and Ali Daud. 'Cloud storage architecture: research challenges and opportunities.' *Computing* 1 (2020): 1.
7. Ilager, Shashikant, Rajeev Muralidhar, and Rajkumar Buyya. 'Artificial Intelligence (AI)-Centric Management of Resources in Modern Distributed Computing Systems'. In *2020 IEEE Cloud Summit*. IEEE (2020). https://doi.org/10.1109/ieeecloudsummit48914.2020.00007.
8. Schmidt, Eric, Bob Work, Safra Catz, Steve Chien, Chris Darby, Kenneth Ford, Jose-Marie Griffiths et al. 'National security commission on artificial intelligence (AI)', *Technical Report* (2021).
9. Prabavathy, S., and I. Ravi Prakash Reddy. 'Exploring the Dichotomy on Opportunities and Challenges of Smart Technologies in Healthcare Systems'. In *Edge-of-Things in Personalized Healthcare Support Systems*, 1–24. Elsevier (2022). https://doi.org/10.1016/b978-0-323-90585-5.00001-1.
10. Matthew, Ugochukwu O., Jazuli S. Kazaure, Amaonwu Onyebuchi, Ogobuchi Okey Daniel, Ibrahim Hassan Muhammed, and Nwamaka U. Okafor. 'Artificial Intelligence Autonomous Unmanned Aerial Vehicle (UAV) System for Remote Sensing in Security Surveillance'. In *2020 IEEE 2nd International Conference on Cyberspac (CYBER NIGERIA)*. IEEE (2021). https://doi.org/10.1109/cybernigeria51635.2021.9428862.

11. Hernandez, L., and A. Castillo. 'Applications of Cloud Computing in Intelligent Vehicles: A Survey'. *Journal of Artificial Intelligence and Machine Learning in Management* 7, no. 1 (2023): 10–24.
12. Benlian, Alexander, William J. Kettinger, Ali Sunyaev, Till J. Winkler, and GUEST EDITORS. 'Special Section: The Transformative Value of Cloud Computing: A Decoupling, Platformization, and Recombination Theoretical Framework'. *Journal of Management Information Systems: JMIS* 35, no. 3 (3 July 2018): 719–39. https://doi.org/10.1080/07421222.2018.1481634.
13. Rajendra, P., Mina Kumari, Sangeeta Rani, Namrata Dogra, Rahul Boadh, Ajay Kumar, and Mamta Dahiya. 'Impact of Artificial Intelligence on Civilization: Future Perspectives'. *Materials Today: Proceedings* 56 (2022): 252–56. https://doi.org/10.1016/j.matpr.2022.01.113.
14. Rani, Sangeeta, Ajay Kumar, Arko Bagchi, Snehlata Yadav, and Sachin Kumar. 'RPL Based Routing Protocols for Load Balancing in IoT Network'. *Journal of Physics. Conference Series* 1950, no. 1 (1 August 2021): 012073. https://doi.org/10.1088/1742-6596/1950/1/012073.
15. Jain, Richa, and Neelam Sharma. 'A Quantum Inspired Hybrid SSA–GWO Algorithm for SLA Based Task Scheduling to Improve QoS Parameter in Cloud Computing'. *Cluster Computing* (29 September 2022). https://doi.org/10.1007/s10586-022-03740-x.
16. Phogat, Manu, Ajay Kumar, Deepak Nandal, and Jyoti Shokhanda. 'A Novel Automating Irrigation Techniques Based on Artificial Neural Network and Fuzzy Logic'. *Journal of Physics Conference Series* 1950, no. 1 (1 August 2021): 012088. https://doi.org/10.1088/1742-6596/1950/1/012088.
17. Keshav, and Sangeeta Rani. 'Big Data Query Processing Approach UsingMongoDB'. In *2023 3rd International Conference on Innovative Practices in Technology and Management (ICIPTM)*. IEEE (2023). https://doi.org/10.1109/iciptm57143.2023.10117738.
18. Rani, Sangeeta, Khushboo Tripathi, Yojna Arora, and Ajay Kumar. 'A Machine Learning Approach to Analyze Cloud Computing Attacks'. In *2022 5th International Conference on Contemporary Computing and Informatics (IC3I)*. IEEE (2022). https://doi.org/10.1109/ic3i56241.2022.10073468.
19. Jain, Richa, and Neelam Sharma. 'A Deadline-Constrained Time-cost-effective Salp Swarm Algorithm for resource optimization in Cloud Computing'. *International Journal of Applied Metaheuristic Computing* 13, no. 1 (20 January 2022): 1–21. https://doi.org/10.4018/ijamc.292509.

Chapter 11

HDVF

Hazy object detection network based on variational features

Kavinder Singh and Anil Singh Parihar

Department of Computer Science & Engineering, Delhi Technological University, New Delhi, India

11.1 INTRODUCTION

In hazy environments, the quality of captured images degrades significantly due to faded details and color distortion, which leads to challenges in object detection. This work focuses on improving the detection performance in hazy conditions. Vision-based applications (e.g., object detection, human action recognition) are not trained for the real-world challenging environment and sometimes fail to generalize in such conditions. Object detection in a hazy environment is required by diverse applications like animal detection, surveillance, and object tracking. Thus, it is crucial to develop an appropriate method to deal with hazy object detection. To deal with hazy conditions, the simplest method is to perform pre-processing using the dehazing or other enhancement algorithms [1–3] prior to detection. Such a method involves computational overhead and additional processing time. Further, the performance of these methods depends on the dehazing algorithm. Researchers developed domain adaptation (DA) and trained the detector on dehazing data to achieve better performance. DA-based methods perform well for the dataset with similar haze distribution as the training dataset; however, fails to perform adequately for other datasets.

This work focuses on the development of a better feature extraction mechanism for hazy images to enhance the object detection performance. The method exerts a single input for feature extraction using the backbone network with pre-training. The extracted features are then processed by the neck module of the detection method. The head mechanism of the method is responsible for final object detection and localization. Most existing methods work on either of the modules to enhance detection performance, but feature extraction from hazy images is a complicated task. Therefore, this work aims to improve feature extraction from hazy images to achieve better detection results in a hazy environment.

The major problem in hazy object detection is the extraction of features from hazy images. Existing object detection methods are developed to deal with images that contain images with clearly visible details. Therefore,

 DOI: 10.1201/9781003502470-11

current techniques are inadequate to obtain efficient characteristics from foggy pictures to execute object recognition. To overcome this issue, we developed a variational image for feature extraction. The variational image is not exactly the dehazed image, since we used only the transmission estimation to divide the input image for feature extraction. The variational image contains a reduced haze density, which facilitates feature extraction from the hazy regions. The features are also extracted from the input hazy image to facilitate feature extraction from the less hazy regions and avoid the issue of noise in the variational image. The features extracted from hazy and variational images are processed using the feature pyramid network. The processed features are used by the detection head to perform effective object detection.

Our approach stands out due to the estimation of variational images for feature extraction. We have also created a model that combines the features of the hazy image and the variational image for hazy object detection. Our method has been proven to be more effective than other existing methods, as shown by quantitative analysis.

11.2 RELATED WORK

The performance improvement is noticeable in various domains [4–10], including object detection after the involvement of artificial intelligence. Two-stage and one-stage approaches are the two broad categories of object detection methods. In the two-stage approaches, the initial phase is dedicated to providing regions proposals, where methods such as Selective Search [11], EdgeBoxes [12], and Deep-Mask [13] are employed. Subsequently, in the second stage, the classification of objects is carried out, accompanied by improved bounding boxes. However, it is important to note that two-stage detectors, such as R-CNNs [14–16] and SPPnet [17], require a comparatively higher computational time and are less computationally efficient compared to one-stage approaches.

Due to improved efficiency, one-stage detectors [18–20] have received considerable attention. YOLO [21, 22] significantly accelerated object detection by presenting an end-to-end model that performs detection through regression. SSD [23] has endeavored to elevate detection performance by incorporating multiple aspect scales and ratios to accommodate objects of varying sizes. Based on the SSD framework, several foundational approaches [24, 25] have been developed for training object detectors from scratch. An inherent challenge in object detection arises from the uneven distribution of samples across different object classes within datasets, resulting in class imbalance. To mitigate this issue, RetinaNet [20] introduced focal loss, effectively managing the weighting of easy negatives and challenging examples. Furthermore, scale-transfer object detection (STOD) [24, 26] introduced scale-transfer techniques to explicitly explore consistency across multiple scales. Local reconfiguration and global attention within the feature pyramid were presented by the deep feature pyramid reconfiguration (DFPR) method [27]. RFBN [28] improved feature discrimination capabilities through the development of receptive field blocks. The aforementioned object detection methods have been designed and trained primarily for normal images. However, their effectiveness is notably degraded when working with images captured in hazy environments.

Object detection in a hazy environment is less explored than general object detection. The enhancement [29–31] of hazy images may help in improving the detection performance. However, such an approach increases the computational and time requirement. Thus, few methods attempted to solve detection tasks in a hazy environment by using domain adaptation. Sindagi et al. [32] developed object detection using a transmission map as the domain before performing domain adaptation. Kumar et al. [33] presented object detection after dehazing the image using a dark channel prior to [1] for autonomous vehicles. Kalwar et al. [34] presented a domain-agnostic network using gated differentiable image processing to plug it into existing object detection models. Liu et al. [35] developed object detection guided by enhancement to use enhancement to refine the detection model. Li et al. [36] presented enhancement of the dehaze image prior to detection. Tanwar et al. [37] presented an image dehazing model for object detection. Qin et al. [38] proposed a detection-driven enhancement network that combines low-frequency and high-frequency components for multiscale features. Li et al. [39] combined the dehazing and detection networks to propose detection-friendly dehazing.

11.3 THE PROPOSED HDVF METHOD

We developed a method for detecting hazy objects based on variational characteristics. The proposed method uses a variational optimization to help the feature extractor in the extraction of haze-relevant features. The details are discussed in the following subsections. The block architecture is shown in Figure 11.1.

11.3.1 Variational module

In hazy images, transmission maps and atmospheric light are the two important factors for the removal of the haze. The transmission map may vary for the overall image as the haze density in an image varies according to the scene depth. The object at higher depth may suffer from more dense haze than the closer objects. We estimate the transmission map using variational optimization to extract haze-relevant features. We estimate the transmission map by maximizing structural details and minimizing textural details. The

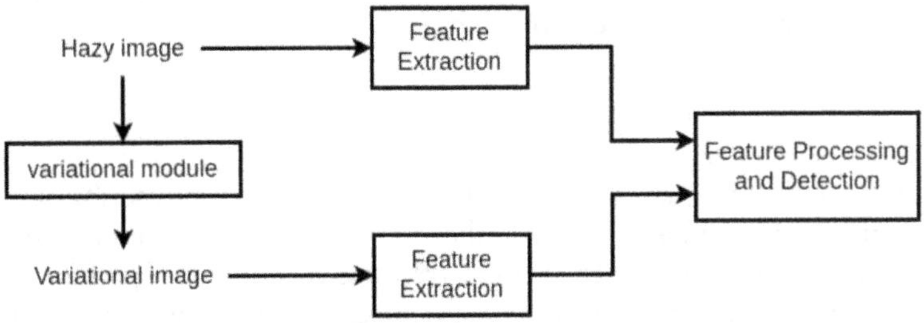

Figure 11.1 Our hazy object detection network based on variational features.

basic estimate of the transmission map uses adaptive bilateral filtering [2]. The adaptive bilateral filtering uses an adaptive scale to perform the textural smoothing. The basic transmission can be expressed as

$$\hat{\boldsymbol{T}} = \partial\left(\max(\boldsymbol{I})\right) \tag{11.1}$$

where $\boldsymbol{I}$ denotes the hazy image, ∂ denotes the adaptive bilateral filtering, $\hat{\boldsymbol{T}}$ denotes the basic transmission, and max(o) denotes the element-wise max operation.

Our variational optimization is a minimization problem that is formulated using the basic transmission and the regularization terms. Out optimization focuses on structural smoothing to improve the transmission map. Our optimization is expressed as

$$\min_{T}\left\|\hat{\boldsymbol{T}} - \boldsymbol{T}\right\|_2^2 + \propto \left\|\nabla \boldsymbol{I}_v - \nabla \boldsymbol{T}\right\|_2^2 + \beta\left\|\boldsymbol{G}^\circ \nabla \boldsymbol{T}\right\|_1 \tag{11.2}$$

where ∇ represents the first-order derivative operator, $\boldsymbol{T}$ is refined transmission, $\nabla \boldsymbol{I}_v$ denotes derivative of *V*-channel of HSV, $\boldsymbol{G}$ is the weight matrix, $\|\nabla\|_p$ is the L_p norm operator. The fidelity among basic and refined transmission is provided in the optimization problem using the first term. The second term in the optimization is used for structural detail maximization. The last term helps in minimizing textural detail and leads to overall smoothing.

The estimated transmission contains characteristics related to the density of the haze. Therefore, it is not feasible to extract the haze-relevant features from the estimated transmission. The atmospheric light in the hazy images varies from scene to scene. We proposed an element-wise division of the transmission map and the input image by ignoring the constraint related to the atmospheric light in the hazy images, termed the variational image. The variational image (V) can be expressed as

$$V = I/T \tag{11.3}$$

The variational image is used as the input to the feature extraction network along with the hazy image. Almost all detectors are trained for clear images, not for variational images; thus, joint training is required to detect objects from the variational image and the hazy image. We estimate the variational image using our variational optimization method and then train the detector on the variational image for object detection in hazy conditions.

11.3.2 Feature extraction module

We developed a feature extraction module that uses both hazy images and variational images. The feature extraction module uses backbone networks for both hazy images and variational images for feature extraction. The backbone of the hazy image extracts all possible features from the hazy images; however, due to limited visibility, it sometimes misses features from the hazy regions. Thus, we developed a backbone with variational images to facilitate feature extraction from the hazy regions. The variational images contain slightly better images with lower haze density. However, the variational images are not haze-free images, since we just used the transmission map while estimating the variational images. Both backbones extract multi-level features from the inputs, which can be expressed as

$$F_i = [B_i^3, B_i^2, B_i^1] \tag{11.4}$$

where $i \in$ [hazy, variational], F_i is the feature pyramid of the ith backbone, and B_i^j denotes the jth level feature from the respective backbone.

11.3.3 Feature processing and detection module

The HDVF uses the features extracted from the hazy and variational images. The feature pyramids are processed using the feature pyramid networks for each set of features. The processed pyramids are used to perform detection and localization. The subnetworks for class detection and box prediction process these features to perform the final detection. The proposed method uses the loss function, class subnet, and box subnet of generalized focal loss [40]. The class and box subnets use distinct sets of convolution layers to process the features from the fused feature pyramid for class and box prediction, which can be represented as

$$F = Conv(Conv(B_l)) \tag{11.5}$$

where, $\forall B_l$ in the respective pyramids. The class and box subnet uses the feature of all the levels from the fused feature pyramid for prediction. The loss function L is expressed as

$$L = -\sum_{m=1}^{c} \alpha_m (1 - p_m^n)^{\gamma} \log(p_m^n) \tag{11.6}$$

where α_m represents weights of mth class, p_m^n represents the probability of class c for sample t, c represents the overall class number, and γ controls the degree of weights of samples.

11.4 EXPERIMENTAL RESULTS AND ANALYSIS

In this section, we will cover the datasets used in our approach, the specifics of the implementation, and the quantitative evaluation. The datasets are the primary need of any deep-leaning method; thus, the discussion of the datasets is given in the following subsection. The details related to the implementation and validation are also discussed to provide all hyper parameters. Finally, superior performance is demonstrated using quantitative analysis.

11.4.1 Dataset

In order to extract effective features from hazy images, the presently available object detection methods use pre-trained models on available datasets such as Pascal VOC [41]. Existing object detection methods for hazy conditions use the domain adaptation method to learn the characteristics of the haze. The method uses the image dehazing mechanism to adapt the characteristics of the haze. Furthermore, the RTTS dataset [42], which is part of the RESIDE dataset, is used to validate object detection algorithms for hazy environments. In these cases, object detection techniques learn haze characteristics from the RESIDE dataset (similar characteristics as the RTTS dataset). Thus, while these methods perform detection in an acceptable manner for the synthetic haze of the RTTS dataset, they fall short in the real-world scenario or other cases. Thus, we synthesize hazy images to train our hazy object detection method using annotated images from the VOC dataset. We can train our method from start to finish and show that it can be applied to the existing dataset. Our HDVF is trained using our synthetic dataset and validated using the benchmarking RTTS dataset. We synthesized the hazy images using the physical image formation model [43,44], which uses a transmission map and atmospheric light. The transmission map is estimated using the depth map by the monodepth [45] technique.

11.4.2 Implementation details

The Pytorch library is utilized for the implementation of the HDVF model. Our HDVF model uses pre-trained ResNet [46] architectures for feature extraction from both hazy and variational images. The training of our HDVF uses a batch size of ten samples per GPU on a dual Titan-RTX GPU machine with Ubuntu OS and 128GB RAM. The training uses a synthetic dataset, and the testing uses the RTTS dataset. The proposed HDVF uses ten epoch training on the synthetic dataset. The optimizer used is SGD (stochastic gradient descent). For all experiments, the learning rate in the SGD optimizer is set to 0.01 without any reduction.

11.4.3 Quantitative analysis

The quantitative analysis of existing contemporary techniques and our method is performed with the help of the RTTS dataset. To ensure the fairness in comparison, the values of the contemporary methods is considered as provided by the authors of the respective methods. The quantitative evaluation is performed with the help of mean average precision and the class average precision with 0.5 as the threshold IoU (Intersection over Union). The other contemporary methods (in quantitative analysis) uses ResNet, and VGG as their backbones. In a hazy environment, the RTTS dataset is mostly used for the analysis of object detection methods.

Table 11.1 demonstrates the performance comparison of our HDVF method and other contemporary techniques on the RTTS dataset. The other contemporary methods use the RESIDE dataset for learning the haze characteristics. RESIDE is a well-known image dehazing dataset that includes over 4000 hazy images and is a superset of RTTS. The RTTS

Table 11.1 Performance analysis of various methods on detection task with GFL detection model

Method	*Person*	*Car*	*Bus*	*Bike*	*Bicycle*	*mAP*
Faster RCNN [16]	46.6	39.8	11.7	19.0	37.0	30.9
DA-Faster [47]	42.5	43.7	16.0	18.3	32.8	30.7
SWDA [48]	40.1	44.2	16.6	23.2	41.3	33.1
SCL [49]	33.5	48.1	18.2	15.0	28.9	28.7
PBDA [32]	37.4	**54.7**	17.2	22.5	38.5	34.1
SADA [50]	37.9	52.7	14.5	16.1	26.2	29.5
DA-ODFS [51]	47.7	53.4	19.1	30.2	**49.3**	39.9
Proposed	**65.4**	54.6	**20.4**	**35.1**	45.5	**44.2**

dataset consists of five classes: bicycle, bike, bus, car, and person. Instead of training on the RTTS dataset, we trained on a synthesized dataset and then validated the model on the RTTS dataset. Our proposed method outperformed other techniques for most classes, although our BFF method achieved the second-best score for the bicycle and car classes. The mAP score shows that our HDVF method convincingly outperforms other techniques.

11.5 CONCLUSION

The chapter presents a variational feature-based hazy object detection method. The HDVF method developed a variational image to help the feature extraction from the hazy images. Our approach entails solving a multi-objective optimization problem to produce a variational image, which is then used to extract variational features that improve object recognition. By incorporating both the original hazy image and the variational image, our method extracts features that are used for object detection and processing using the feature processing and detection network. The proposed method uses the synthetic dataset for training. We evaluated our method on the RTTS dataset, which demonstrates the superior performance of the proposed method in hazy conditions.

REFERENCES

[1] He, Kaiming, Jian Sun, and Xiaoou Tang. "Single image haze removal using dark channel prior." *IEEE Transactions on Pattern Analysis and Machine Intelligence* 33, no. 12 (2010): 2341–2353. https://doi.org/10.1109/TPAMI.2010.168

[2] Ghosh, Sanjay, Ruturaj G. Gavaskar, Debasisha Panda, and Kunal N. Chaudhury. "Fast scale-adaptive bilateral texture smoothing." *IEEE Transactions on Circuits and Systems for Video Technology* 30, no. 7 (2019): 2015–2026.

[3] Parihar, Anil Singh, Kavinder Singh, Hrithik Rohilla, Gul Asnani, and Harpreet Kour. "A comprehensive analysis of fusion-based image enhancement techniques." In *2020 4th International Conference on Intelligent Computing and Control Systems (ICICCS)*, pp. 823–828. IEEE, 2020. https://doi.org/10.1109/ICICCS48265.2020.9120999

[4] Rajendra, P., Mina Kumari, Sangeeta Rani, Namrata Dogra, Rahul Boadh, Ajay Kumar, and Mamta Dahiya. "Impact of artificial intelligence on civilization: Future perspectives." *Materials Today: Proceedings* 56 (2022): 252–256.

[5] Rani, Sangeeta, Ajay Kumar, Arko Bagchi, Snehlata Yadav, and Sachin Kumar. "RPL based routing protocols for load balancing in IoT network." *Journal of Physics: Conference Series*, 1950, no. 1 (2021), 012073.

[6] Phogat, Manu, Ajay Kumar, Deepak Nandal, and Jyoti Shokhanda. "A novel automating irrigation techniques based on artificial neural network

and fuzzy logic." *Journal of Physics: Conference Series*, 1950, no. 1 (2021), 012088.

[7] Rani, Sangeeta, Khushboo Tripathi, Yojna Arora, and Ajay Kumar. "Analysis of anomaly detection of Malware using KNN." In *2022 2nd International Conference on Innovative Practices in Technology and Management (ICIPTM)*, vol. 2, pp. 774–779. IEEE (2022). https://doi.org/10.1109/ICIPTM54933.2022.9754044

[8] Rani, Sangeeta, Khushboo Tripathi, Yojna Arora, and Ajay Kumar. "A machine learning approach to analyze cloud computing attacks." In *2022 5th International Conference on Contemporary Computing and Informatics (IC3I)*, pp. 22–26. IEEE (2022). https://doi.org/10.1109/IC3I56241.2022.10073468

[9] Kumar, Ajay, Hari Singh, Parveen Kumar, and Bandar AlMangour, eds. *Handbook of Smart Manufacturing: Forecasting the Future of Industry 4.0*. CRC Press (2023).

[10] Sethi, Angad, and Kavinder Singh. "Natural language processing based automated essay scoring with parameter-efficient transformer approach." In *2022 6th International Conference on Computing Methodologies and Communication (ICCMC)*, pp. 749–756. IEEE (2022).

[11] Uijlings, Jasper RR, Koen EA Van De Sande, Theo Gevers, and Arnold WM Smeulders. "Selective search for object recognition." *International Journal of Computer Vision,* 104 (2013): 154–171.

[12] Zitnick, C. Lawrence, and Piotr Dollár. "Edge boxes: Locating object proposals from edges." In *Computer Vision–ECCV 2014: 13th European Conference, Zurich, Switzerland, September 6–12, 2014, Proceedings, Part V 13*, pp. 391–405. Springer International Publishing (2014).

[13] O Pinheiro, Pedro O., Ronan Collobert, and Piotr Dollár. "Learning to segment object candidates." Advances in *Neural Information Processing Systems* 28 (2015): 1–9.

[14] Girshick, Ross, Jeff Donahue, Trevor Darrell, and Jitendra Malik. "Rich feature hierarchies for accurate object detection and semantic segmentation." In *Proceedings of the IEEE Conference on Computer Vision and Pattern Recognition*, pp. 580–587, 2014. https://doi.org/10.1109/CVPR.2014.81

[15] Girshick, Ross. "Fast r-cnn." In *Proceedings of the IEEE International Conference on Computer Vision*, pp. 1440–1448, 2015. https://doi.org/10.1109/ICCV.2015.169

[16] Ren, Shaoqing, Kaiming He, Ross Girshick, and Jian Sun. "Faster r-cnn: Towards real-time object detection with region proposal networks." *Advances in Neural Information Processing Systems* 28 (2015), 1137–1149.

[17] He, Kaiming, Xiangyu Zhang, Shaoqing Ren, and Jian Sun. "Spatial pyramid pooling in deep convolutional networks for visual recognition." *IEEE Transactions on Pattern Analysis and Machine Intelligence* 37, no. 9 (2015): 1904–1916. https://doi.org/10.1109/TPAMI.2015.2389824

[18] Redmon, Joseph, Santosh Divvala, Ross Girshick, and Ali Farhadi. "You only look once: Unified, real-time object detection." In *Proceedings of the IEEE Conference on Computer Vision and Pattern Recognition*, pp. 779–788. 2016.

[19] Redmon, Joseph, and Ali Farhadi. "YOLO9000: Better, faster, stronger." In *Proceedings of the IEEE Conference on Computer Vision and Pattern Recognition*, pp. 7263–7271 (2017).
[20] Lin, Tsung-Yi, Priya Goyal, Ross Girshick, Kaiming He, and Piotr Dollár. "Focal loss for dense object detection." In *Proceedings of the IEEE International Conference on Computer Vision*, pp. 2980–2988 (2017).
[21] Redmon, Joseph, and Ali Farhadi. "Yolov3: An incremental improvement." arXiv preprint arXiv:1804.02767 (2018).
[22] Bochkovskiy, Alexey, Chien-Yao Wang, and Hong-Yuan Mark Liao. "Yolov4: Optimal speed and accuracy of object detection." arXiv preprint arXiv:2004.10934 (2020).
[23] Liu, Wei, Dragomir Anguelov, Dumitru Erhan, Christian Szegedy, Scott Reed, Cheng-Yang Fu, and Alexander C. Berg. "SSD: Single shot multibox detector." In *Computer Vision–ECCV 2016: 14th European Conference, Amsterdam, The Netherlands, October 11–14, 2016, Proceedings, Part I 14*, pp. 21–37. Springer International Publishing (2016).
[24] Shen, Zhiqiang, Zhuang Liu, Jianguo Li, Yu-Gang Jiang, Yurong Chen, and Xiangyang Xue. "Dsod: Learning deeply supervised object detectors from scratch." In *Proceedings of the IEEE International Conference on Computer Vision*, pp. 1919–1927 (2017). https://doi.org/10.1109/ICCV.2017.212
[25] Zhu, Chenchen, Yihui He, and Marios Savvides. "Feature selective anchor-free module for single-shot object detection." In *Proceedings of the IEEE/CVF Conference on Computer Vision and Pattern Recognition*, pp. 840–849 (2019).
[26] Zhu, Rui, Shifeng Zhang, Xiaobo Wang, Longyin Wen, Hailin Shi, Liefeng Bo, and Tao Mei. "ScratchDet: Training single-shot object detectors from scratch." In *Proceedings of the IEEE/CVF Conference on Computer Vision and Pattern Recognition*, pp. 2268–2277 (2019). https://doi.org/10.1109/CVPR.2019.00237
[27] Kong, Tao, Fuchun Sun, Chuanqi Tan, Huaping Liu, and Wenbing Huang. "Deep feature pyramid reconfiguration for object detection." In *Proceedings of the European Conference on Computer Vision (ECCV)*, pp. 169–185 (2018).
[28] Liu, Songtao, and Di Huang. "Receptive field block net for accurate and fast object detection." In *Proceedings of the European Conference on Computer Vision (ECCV)*, pp. 385–400 (2018).
[29] Ghosh, Sanjay, and Kunal N. Chaudhury. "Fast bright-pass bilateral filtering for low-light enhancement." In *2019 IEEE International Conference on Image Processing (ICIP)*, pp. 205–209. IEEE (2019).
[30] Singh, Kavinder, and Anil Singh Parihar. "DSE-Net: Deep simultaneous estimation network for low-light image enhancement." *Journal of Visual Communication and Image Representation* 91 (2023): 103780.
[31] Singh, Kavinder, and Anil Singh Parihar. "Illumination Estimation for Nature Preserving low-light image enhancement." *The Visual Computer* (2023): 121–136.
[32] Sindagi, Vishwanath A., Poojan Oza, Rajeev Yasarla, and Vishal M. Patel. "Prior-based domain adaptive object detection for hazy and rainy

conditions." In *Computer Vision–ECCV 2020: 16th European Conference, Glasgow, UK, August 23–28, 2020, Proceedings, Part XIV 16*, pp. 763–780. Springer International Publishing (2020).

[33] Kumar, Bagesh, Amritansh Mishra, Aayush Talesara, Shubham Kumar, Sourav Dey, Vyas, and Ranjana Vyas. "Object detection for autonomous vehicle in Hazy environment using optimized deep learning techniques." In *Proceedings of the 2022 Fourteenth International Conference on Contemporary Computing*, pp. 242–249 (2022).

[34] Kalwar, Sanket, Dhruv Patel, Aakash Aanegola, Krishna Reddy Konda, Sourav Garg, and K. Madhava Krishna. "GDIP: gated differentiable image processing for object detection in adverse conditions." In *2023 IEEE International Conference on Robotics and Automation (ICRA)*, pp. 7083–7089. IEEE (2023).

[35] Liu, Wenyu, Gaofeng Ren, Runsheng Yu, Shi Guo, Jianke Zhu, and Lei Zhang. "Image-adaptive YOLO for object detection in adverse weather conditions." *Proceedings of the AAAI Conference on Artificial Intelligence*, 36, no. 2 (2022): 1792–1800.

[36] Li, Binghan, Yindong Hua, and Mi Lu. "Object detection in Hazy environment enhanced by preprocessing image dataset with synthetic haze." In *2020 International Conference on Computational Science and Computational Intelligence (CSCI)*, pp. 1618–1623. IEEE (2020).

[37] Tanwar, Ritik, Shubham Verma, and Manoj Kumar. "Object detection using image dehazing: A journey of visual improvement." In *2022 2nd International Conference on Intelligent Technologies (CONIT)*, pp. 1–8. IEEE (2022).

[38] Qin, Qingpao, Kan Chang, Mengyuan Huang, and Guiqing Li. "DENet: Detection-driven enhancement network for object detection under adverse weather conditions." In *Proceedings of the Asian Conference on Computer Vision*, pp. 2813–2829 (2022).

[39] Li, Chengyang, Heng Zhou, Yang Liu, Caidong Yang, Yongqiang Xie, Zhongbo Li, and Liping Zhu. "Detection-friendly dehazing: Object detection in real-world hazy scenes." *IEEE Transactions on Pattern Analysis and Machine Intelligence* 45, no. 7 (2023): 8284–8295.

[40] Li, Xiang, Chengqi Lv, Wenhai Wang, Gang Li, Lingfeng Yang, and Jian Yang. "Generalized focal loss: Towards efficient representation learning for dense object detection." *IEEE Transactions on Pattern Analysis and Machine Intelligence* 45, no. 3 (2022): 3139–3153. https://doi.org/10.1109/TPAMI.2022.3180392

[41] Everingham, Mark, Luc Van Gool, Christopher KI Williams, John Winn, and Andrew Zisserman. "The pascal visual object classes (VOC) challenge." *International Journal of Computer Vision* 88 (2010): 303–338.

[42] Li, Boyi, Wenqi Ren, Dengpan Fu, Dacheng Tao, Dan Feng, Wenjun Zeng, and Zhangyang Wang. "Benchmarking single-image dehazing and beyond." *IEEE Transactions on Image Processing* 28, no. 1 (2018): 492–505.

[43] Nayar, Shree K., and Srinivasa G. Narasimhan. "Vision in bad weather." In *Proceedings of the Seventh IEEE International Conference on Computer Vision*, vol. 2, pp. 820–827. IEEE (1999). https://doi.org/10.1109/ICCV.1999.790306

[44] Narasimhan, Srinivasa G., and Shree K. Nayar. "Chromatic framework for vision in bad weather." In *Proceedings IEEE Conference on Computer Vision and Pattern Recognition. CVPR 2000 (Cat. No. PR00662)*, vol. 1, pp. 598–605. IEEE (2000).

[45] Godard, Clément, Oisin Mac Aodha, Michael Firman, and Gabriel J. Brostow. "Digging into self-supervised monocular depth estimation." In *Proceedings of the IEEE/CVF International Conference on Computer Vision*, pp. 3828–3838 (2019). https://doi.org/10.1109/ICCV.2019.00393

[46] He, Kaiming, Xiangyu Zhang, Shaoqing Ren, and Jian Sun. "Deep residual learning for image recognition." In *Proceedings of the IEEE Conference on Computer Vision and Pattern Recognition*, pp. 770–778 (2016).

[47] Chen, Yuhua, Wen Li, Christos Sakaridis, Dengxin Dai, and Luc Van Gool. "Domain adaptive faster r-cnn for object detection in the wild." In *Proceedings of the IEEE Conference on Computer Vision and Pattern Recognition*, pp. 3339–3348 (2018).

[48] Saito, Kuniaki, Yoshitaka Ushiku, Tatsuya Harada, and Kate Saenko. "Strong-weak distribution alignment for adaptive object detection." In *Proceedings of the IEEE/CVF Conference on Computer Vision and Pattern Recognition*, pp. 6956–6965 (2019).

[49] Shen, Zhiqiang, Harsh Maheshwari, Weichen Yao, and Marios Savvides. "Scl: Towards accurate domain adaptive object detection via gradient detach based stacked complementary losses." arXiv preprint arXiv:1911.02559 (2019).

[50] Chen, Yuhua, Haoran Wang, Wen Li, Christos Sakaridis, Dengxin Dai, and Luc Van Gool. "Scale-aware domain adaptive faster r-cnn." *International Journal of Computer Vision* 129, no. 7 (2021): 2223–2243.

[51] Yang, Xin, Michael Bi Mi, Yuan Yuan, Xin Wang, and Robby T. Tan. "Object Detection in Foggy scenes by embedding depth and reconstruction into domain adaptation." In *Proceedings of the Asian Conference on Computer Vision*, pp. 1093–1108 (2022).

Index

Note: Page numbers in **bold** refer to tables and *italics* refer to figures.

For Product Safety Concerns and Information please contact our EU
representative GPSR@taylorandfrancis.com
Taylor & Francis Verlag GmbH, Kaufingerstraße 24, 80331 München, Germany

www.ingramcontent.com/pod-product-compliance
Lightning Source LLC
LaVergne TN
LVHW020638300926
849161LV00004B/748

* 9 7 8 1 0 3 2 8 0 1 2 8 5 *